NEW RESEARCH ON THE ENVIRONMENT AND BIOTECHNOLOGY

NEW RESEARCH ON THE ENVIRONMENT AND BIOTECHNOLOGY

A. E. KUZNETSOV

AND

G. E. ZAIKOV

EDITORS

Nova Science Publishers, Inc.

New York

For permission to use material from this book please contact us:
Telephone 631-231-7269; Fax 631-231-8175
Web Site: http://www.novapublishers.com

LIBRARY OF CONGRESS CATALOGING-IN-PUBLICATION DATA

New research on the environment and biotechnology / A.E. Kuznetsov and G.E. Zaikov (editors).
p. ; cm.
Includes bibliographical references and index.
ISBN 1-60021-042-2
1. Bioremediation. 2. Hydrocarbons--Biodegradation.
[DNLM: 1. Biodegradation. 2. Biotechnology. 3. Environmental Pollutants--metabolism. WA 671 N532 2006] I. Zaikov, Gennadii Efremovich. II. Kuznetsov, A. E. (Aleksandr Evgen'evich), 1958-
TD192.5.N49 2006
628.5--dc22 2006003808

Published by Nova Science Publishers, Inc. ✢ New York

CONTENTS

In: New Research on the Environment and Biotechnology ISBN 1-60021-042-2
Editors: A. Kuznetsov and G. Zaikov, pp. 1-15

Chapter 1

METAL ACCUMULATION ABILITY OF GRASS SPECIES OF KAZAKHSTAN FLORA AND THE POSSIBILITY OF USING THEM FOR PHYTOREMEDIATION

S. D. Atabayeva, B. A. Sarsenbayev, Sh. K. Kenzhebayeva, B. N. Usenbekov, S. Sh. Asrandina, E. A. Kirshibayev, E. L. Kalmykov*, and A. N. Danilova**

Institute of Plant Physiology, Genetics and Bioengineering, 480090 Almaty, Timiryazeva, 45. E-mail: sauleat@yandex.ru. Fax: 7-(3272)-479292.

* OJSC "Biomedpreparat".474456 Ridder,Microdistrict 9, B. 3

** Altay Botanic Garden, 433910 Leninogorsk, Ermakova, 1

ABSTRACT

The study of grass species in hydroponic conditions indicates that *A. repens* and *S. viridis* are the most tolerant species to Pb and Zn, *Ph. pratense* - most sensitive. The shoot Pb/root Pb ratio was < 1 for all species , but shoot Zn/root Zn ratio was > 1 for species except *A. alba* and *Ag. repens.* The screening of grass species around metallurgic plants of East Kasakhstan have shown that *Dactylis glomerata, Bromus inermis, Phleum pratense* have accumulated Zn and Pb in most amounts mainly in the roots and can be the candidates for using them for phytoimmobilization. *A.repens* and *A. alba* can be used as a cattle forage because of low accumulation of heavy metals in the shoots.

Key words: lead, zinc, phytoremediation, hyperaccumulators, phytoimmobolization.

INTRODUCTION

Large areas of Kazakhstan around metallurgical enterprises are contaminated by heavy metals (HM), especially in the East, South and Central Kazakhstan regions. The mined and processed ores contain, apart from ferrous and non-ferrous metals as lead (Pb), zinc (Zn) and copper (Cu), accompanied by toxic elements – such as arsenic (As), selenium (Se), cadmium (Cd), and radionuclides. Smelting and mining processes have caused all environmental components to be contaminated with a large number of toxic elements. As a consequence, these pollutants also migrate into the food chain. For instance, concentrations of Pb, Zn and Cd are reported to be very large in plants growing in the proximity of metallurgical enterprises of non-ferrous metallurgy. In these areas, the contents of HM in foodstuffs exceed the permissible concentrations. Panin (1998) [1] reported concentration of Pb in grains 5 times larger, and Zn and Pb concentrations in potatoes 5-7 times larger than the allowable limits. Also the concentration of Pb in milk and Pb and Cd in internal organs of livestock was exceeded by 6-12 times and 5-14 times respectively. In humans the concentration of Cd in blood and hairs was 6-8 and 3 times larger than the normal range [1]. Due to the abundance of heavy metals in soils, plants tend to accumulate these elements in shoots and roots. Heavy metals can have toxic effects on plants such as suppression of growth and biomass accumulation, chlorosis, disturbance of the water status, and as consequence, the reducing in crop yield. Therefore it is necessary to find the ways of decrease of plant and soil contamination by heavy metals.

Plants have different responses to toxic metals. One of these is represented by hyperaccumulator plants. These plants accumulate in their shoots metal concentrations exceeding those of normal plants by 100 to 1000 without showing toxicity symptoms [2-7]. Thresholds for hyperaccumulation are: >10 000 for Zn, > 1000 for Cu, Pb, Ni and >100 mg/kg for Cd [6].

Hyperaccumulator plants offer an outstanding opportunity to achieve economic phytoextraction to decontaminate polluted soils. Chemical and physical methods of removing metals from soils cannot be as selective as plant roots, because the plants use ion carriers to selectively accumulate and transport specific metals to their shoots. Moreover, traditional chemical and physical remediation technologies are expensive and likely to generate secondary wastes [2].

There are five fundamental processes by which plants can be used to remediate contaminated soils: *phytoimmobilization (phytostabilization), phytoextraction, phyto-degradation, phytovolitalization, rhizofiltration* [3]. In relation to our research we are interested in two processes: *phytoextraction* and *phytoimmobilization.*

Phytoextraction processes extract both metallic and organic constituents from soil by direct uptake into plants and translocation to aboveground biomass [4]. *Phytoimmobilization* processes prevent the movement and transport of dissolved contaminants using plants to decrease the mobility of pollutants in soils. It is defined as a containment technology with the aim of immobilizing contaminants in the root zone. Basically, three direct plant based immobilization mechanisms, i.e., root adsorption /absorption, plant assisted precipitation/formation of insoluble compounds and alteration of soil factors that affect pollutant mobility may be involved. In addition, plants may indirectly enhance pollutant immobilization through stimulation of microbial fixation through adsorption and uptake, and microbial synthesis of less mobile compounds that ultimately results in incorporation into soil humus [3].

The basic aim of our research was determination of grass species from Kazakhstan flora which able to hyperaccumulate heavy metals.

MATERIAL AND METHODS

Screening of Grass Species

The most wide-spread and typical plants, which are growing in East Kazakhstan near contaminated sites, were taken with roots. Soil from root zones was removed mechanically, without their washings by water considering that washings of roots would be resulted the uneven loss of certain amount of heavy metals from cell walls and between cell space of plants. All samples were processing equally before their analysis on the heavy metal contents. Plant were dried on the air, were divided into roots and shoots. Then they were dried at temperature 105^0 C, were ground and analysed.

Soil samples were taken from the same area where plants were collected, on the depth 15-20 centimetres.

Soil and plant samples were digested in the mixture of HNO_3 and HCl at the temperature 95 ± 5^oC according to standards for operation procedures (SOD). Concentrations of metals in plants and soils were measured by atomic absorption spectrophotometer (Perkin Elmer).

Hydroponic Short-Term Experiments

Seeds of *Agropyron repens, Agrostis alba, Setaria viridis, Bromus inermis* and *Phleum pratense* were sterilized by treatment with 16% H_2O_2 followed by distilled water rinses. Seed were grown on distilled water 7 days, then the seedlings were transferred to solutions of modified Hoaglands medium. On 15-th day the seedlings were transferred to Hoagland medium, containing various concentrations of Pb (450, 900 mg/l) and Zn (350, 700 mg/l) and grew there 6 days. 21-days seedlings were taken for analysis.

RESULTS AND DISCUSSION

1. The Estimation of Heavy Metal Accumulation Ability of Grass Species of Kazakhstan Flora

The species growing around Zinc Plant (ZP) and Lead Plant (LP) in East Kazakhstan were collected and identified. *Dactylis glomerata, Bromus inermis, Agropyron repens, Agrostis alba* and *Phleum pratense* are wide spread species in this region. The level of contamination of soils and plants by heavy metals (HM) was determined. The soils and plants around metallurgical enterprises accumulated relatively low amounts of Cd and Cu (Fig. 1). The concentrations of Pb and Zn in the soils and plant parts were much more than of Cd and Cu. Therefore we emphasized our attention on the accumulation of Zn and Pb by soils and plants.

The soils around Zinc- and Lead-Plants differ in the level of HM contamination. The soils contained Pb from 12,3 up to 12672, 4 mg/kg. The lowest concentration of Pb was found on the west of ZP, the most amount – on the Center of LP. Pb concentrations in the soils is increased in the order: ZP, West < ZP, East < LP, West < LP, South < LP, North < ZP, South

< LP, Center. Zn content in the soils was from 43,2 up to 129792,3 mg/kg. Concentration of Zn was the lowest on the West of ZP. Maximum of Zn content was found on the South of ZP. Concentrations of Zn increased in the order : ZP, West < ZP, East < LP, West < LP, South < LD, East < ZP, Center < LP, North < LP, Center < ZP, North < ZP, South (Fig.1). In this order we compared the species on the HM accumulation ability. The HM concentrations in plants increased with the increasing of HM content in the soils. Plant species accumulated great amounts of Pb and Zn in their parts (Fig. 2-10).

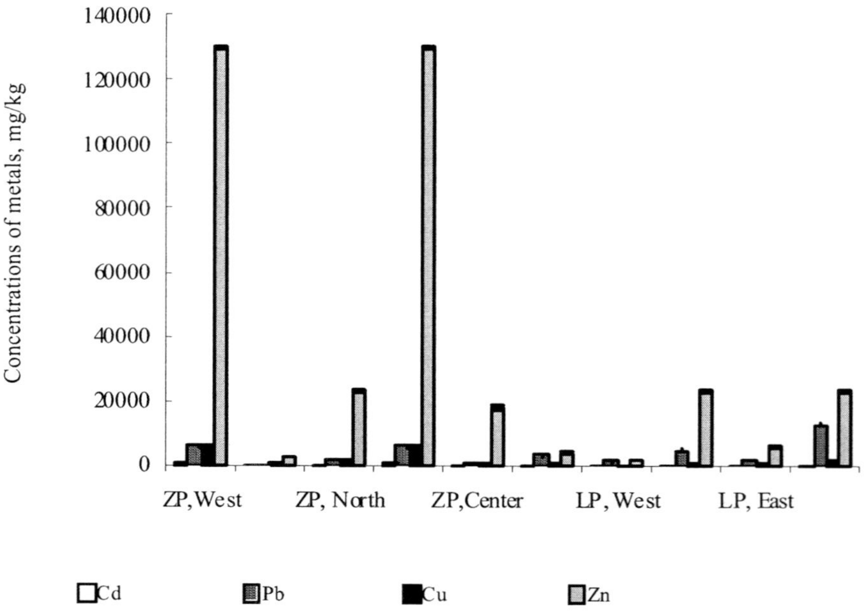

Figure 1. Heavy metals content in the soil

Dactylis glomerata is distinguished by positive correlation of Pb and Zn content in plant part and soil content, especially on strong contaminated soils. With increasing of soil content of Pb and Zn the share of HM in the shoots (LP Center, ZP South) increased. Internal concentrations of Pb (in the roots and shoots) in this plant reflects external Pb levels in the soils (Fig. 2, 7). Shoot/root ratio for Pb was > 1 on the Center and South of ZP.

Bomus inermis accumulated Pb and Zn mainly in the roots. It accumulated the most amount of Pb and Zn on high contaminated soils on the Center of LP (6715,9 mg Pb/kg) and on the North of ZP (4588,9 mg Zn/kg), accordingly (Fig. 3, 8). Shoot/root ratio for Pb on the Center and South of ZP and on the North of LP was > 1, probably because of air contamination. It was observed the considerable amounts of Zn in the shoots on the strong contaminated soils. The shoot/root ration for Zn was more than for Pb, but was < 1.

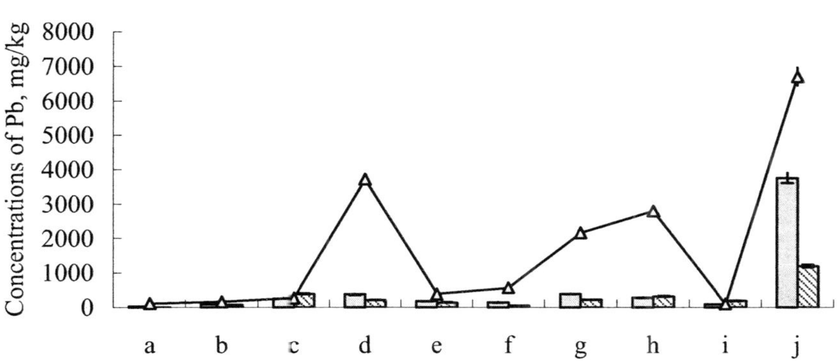

Total soil Pb (mg/kg): a - 12,3; b - 195,8 ; c - 932,8; d - 1308,9; e - 1645,3; f - 1723,9; g - 3046,0 ; h - 4769,2; i - 5990,4; j - 12672,4.

Shoots Ears Roots

Figure 2. Pb content in Dactylis glomerata

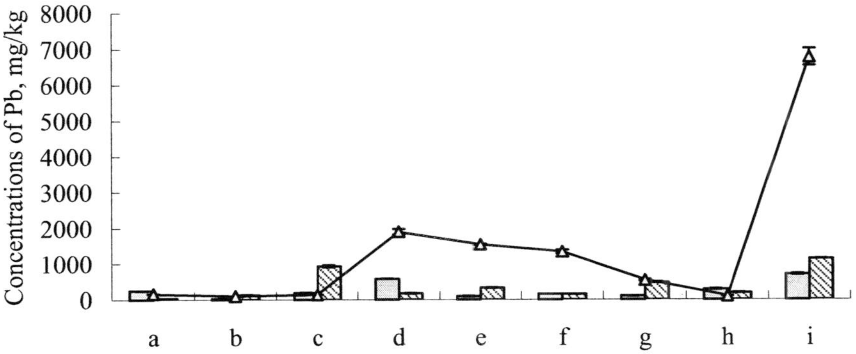

Total soil Pb (mg/kg): a - 12,3; b -195,8 ; c - 932,8; d - 1645,3; e - 1723,9; f - 3046,0; h - 4769,2; g - 5990,4; i - 12672,4.

Shoots Ears Roots

Figure 3. Pb content in Bromus

Agropyron repens is distinguished by low accumulation of Pb in the shoots on the slightly and moderately contaminated soils (Fig. 4). The Pb concentration in the shoots on strong contaminated soils was high, but lower than in the other species. On Pb accumulation by roots it had an intermediate position.

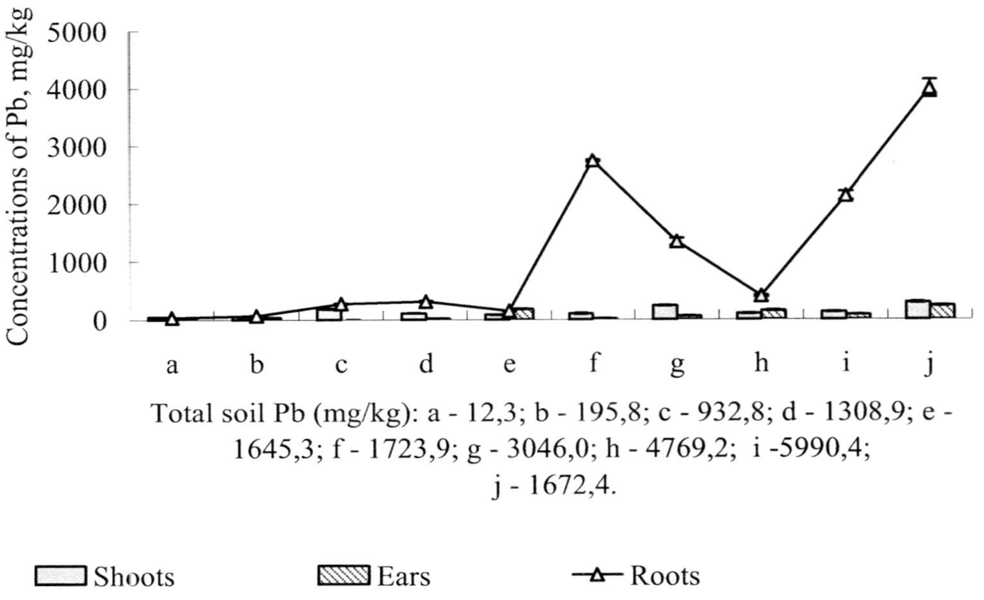

Figure 4. Pb content in Agropyron repens

Concentrations of Zn was lowest in *Agropyron repens* at low or moderate Zn concentrations in the soils (42,3 –3893,0 mg/kg) (Fig. 9). *Agropyron repens* accumulated Zn mainly in the roots, but the translocation rate of Zn from roots to the shoots was higher than of Pb.

Phleum pratense accumulated HM mainly in the roots. It accumulated low amounts of Pb and Zn in the shoots and roots on the soils with low or moderate content of HM in the soils. But on the soils with high concentrations of Pb and Zn *Phleum pratense* accumulated HM in great amounts in the roots.

Thus, the concentrations of Zn and Pb in the shoots and roots of *Dactylis glomerata, Bromus inermis, Agrostis alba* increased with increasing the content of these metals in soil. The Pb content in the shoots of *Agropyron repens* was lowest in comparison with other species and this kind of plant had an intermediate position on Pb accumulation in the roots. On the contrary, *Bromus inermis* accumulated in the roots most amount of Pb. *Dactylis glomerata* accumulated most amounts of Pb in the shoots and roots on strong contaminated soils. Other species are placed between them. With regard to Zn accumulation we can indicate that *Bromus inermis* accumulated most amounts of this metal in the shoots and roots in comparison with other species on high contaminated soils. *Phleum pratense* accumulated the lowest amount of Zn in the shoots on high contaminated soils. *Agrostis alba, Bromus inermis* and *Dactylis glomerata* distinguished by high translocation rate of Zn from roots to shoots on strong contaminated soils.

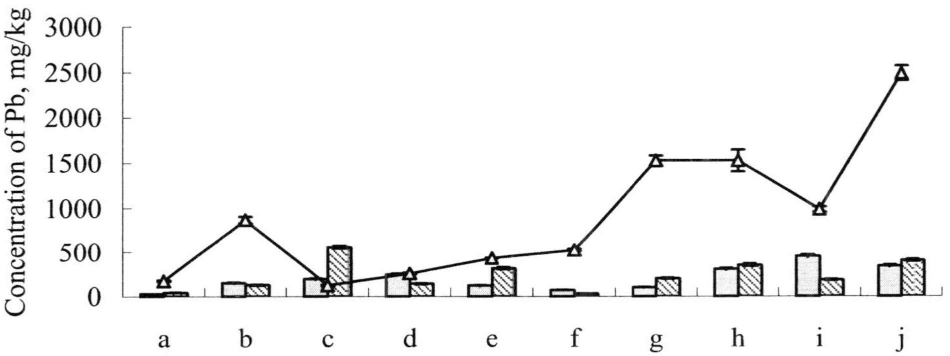

Total soil Pb (mg/kg): a - 12,3; b -195,8; c - 932,8; d - 1308,9; e - 11645,3; f - 1723,9; g - 3046,0; h - 4769,2; i - 5990,4; j - 12672,4.

Shoots   Ears   Roots

Figure 5. Pb content in Agrostis alba

According to literature data a Zn hyperaccumulator is defined as a plant that contains > 10 000 mg of Zn per kg of dry weight, whereas a Pb hyperaccumulator is defined as a plant that contains > 1 000 mg of Pb/kg dry wt. The means of Zn, Pb concentrations expected in nonaccumulating plants growing on contaminated soils are < 1000 for Zn and < 100 for Pb (Brooks, 1998).

Bromus inermis had Pb values in the aboveground parts higher than 1000 mg/kg in the Center of the ZP (Fig. 3). The ratio shoot/root for Pb was 1,2 and in the South and West of ZP was >1. It is probably not a natural phenomenon, because on the other sites we did not observe the hyperaccumulation feature of this plant. On the Center of LP *Bromus inermis* accumulated Pb in the shoots > 1000 mg/kg. The ratio shoot Pb/root Pb was < 1. On the Center of LP *Dactylis glomerata* accumulated in the leaves and ears 3760,0 and 1194,1 mg Pb/kg, accordingly, but the shoot/root ratio was < 1.

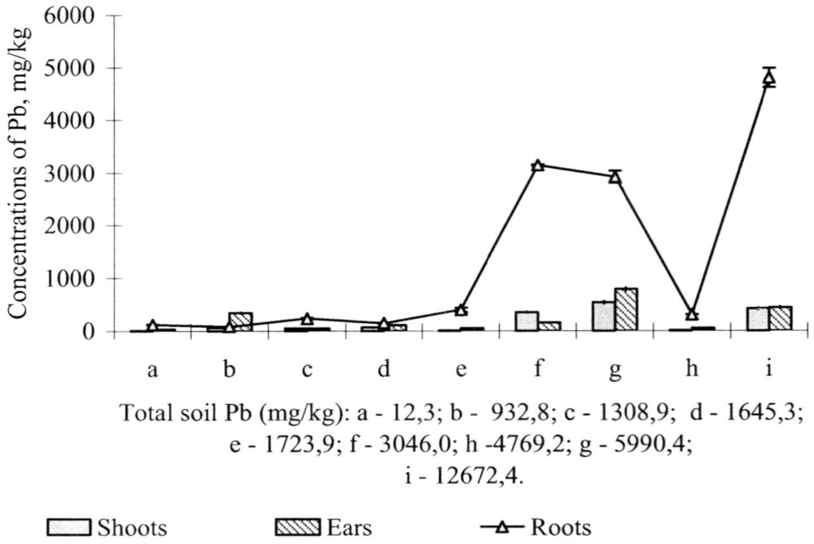

Figure 6. Pb content in Pheum pratense

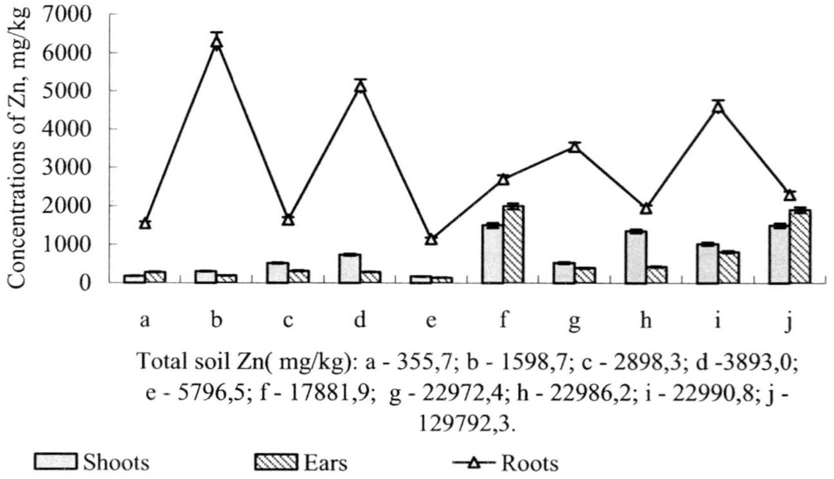

Figure 7. Zn content in Dactylis glomerata

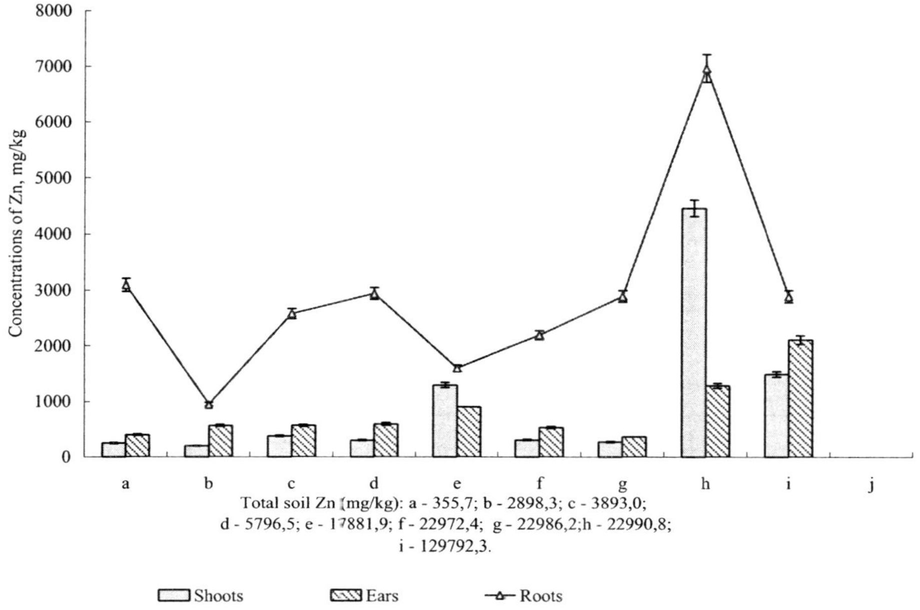

Figure 8. Zn content in Bromus inermis

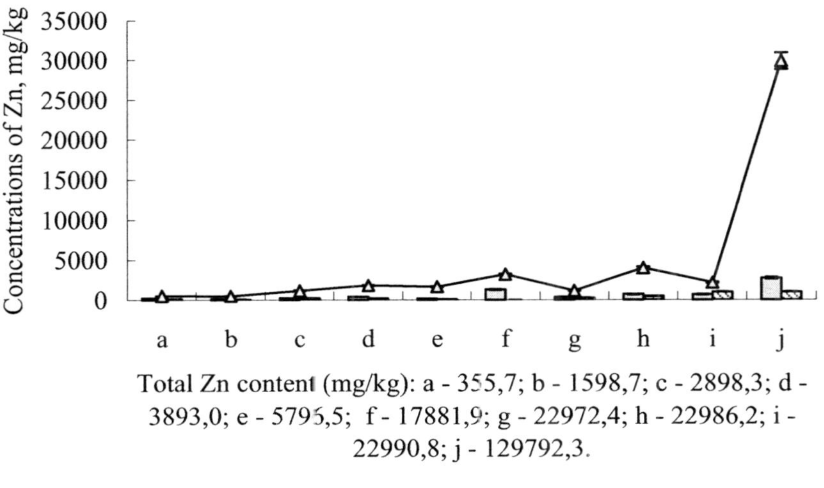

Figure 9. Zn content in Agropyron repens

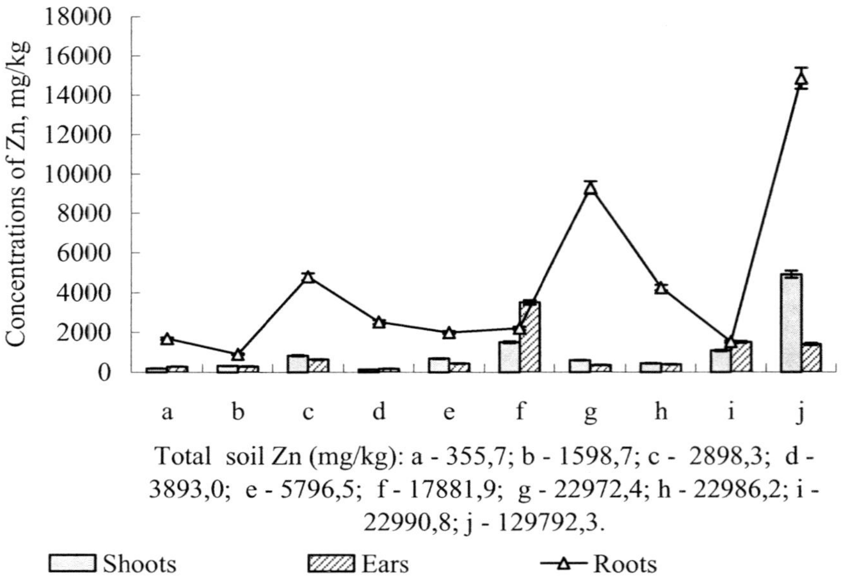

Figure 10. Zn content in Agrostis alba

Thus, we have not found the plants which have all peculiarities of hyperaccumulators. But the content of Pb and Zn for these species was much higher than those defined for non-accumulators growing on contaminated soils. The average concentrations of Zn and Pb expected in nonaccumulating plants growing on contaminated soils are < 1000 for Zn and < 100 for Pb [6]. The average concentrations of Pb for *Dactylis glomerata, Bromus inermis, Agrostis alba* were much above 100 mg/kg (Fig 2-6) and the average concentrations of Zn (Fig. 7-10) for all species on high contaminated soils > 1000. These results require the further study of the accumulation ability of species.

As it is known from literature data two basic strategies of plant response are suggested, *accumulators* and *excluders*. In the former, metals can be concentrated in plant parts from low or high background levels. By contrast, differential uptake and transport between root and shoot in *excluders,* lead to more or less constant levels over a wide range of external concentration. "Indicators" are seen as another mode of response where proportional relationships exist between metal levels in the soil, and uptake and accumulation in plant parts [5]. Metal level in the tissues of indicator species generally reflect metal content in the soils [8]. The metal concentrations in the shoots and roots of *Dactylis glomerata* proportionally increased by increasing metal content in the soils. In our opinion *Dactylis glomerata* belong to *indicators*.

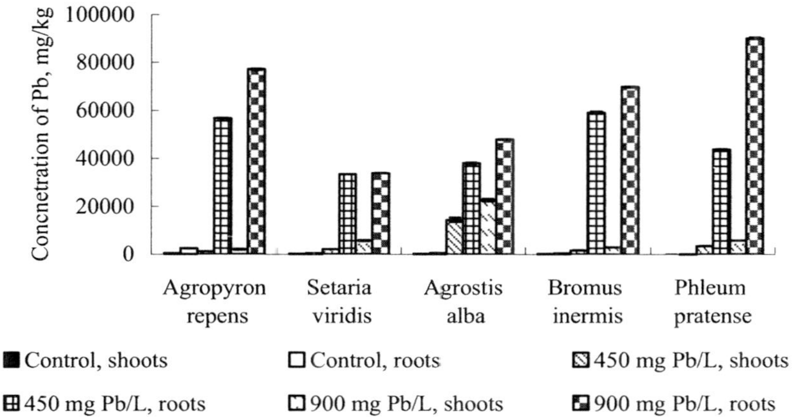

Figure 11. Pb content in plant parts

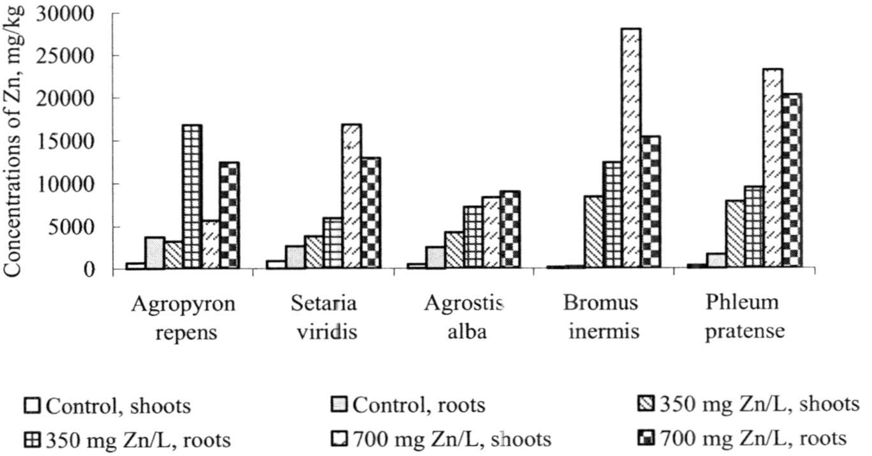

Figure 12. Zn content in plant parts

Metal excluders effectively prevent penetration of metals into aerial parts over a broad range of metal concentrations in the soil, however, they can still contain large amounts of metal in their roots. Therefore *Agropyron repes* can be a preliminarily attribute to *excluders*. As excluders do not in general regulate metal uptake, restriction of transport from roots to shoot is the likely mechanism of control [9].

Agropyron repens, Phleum pratense and *Agrostis alba* can be used on the slightly and moderate contaminated soils as a cattle forage because of low accumulation of heavy metals in the shoots.

In accumulator species root uptake and transport are more or less in balance but metals can accumulate in the roots [5]. It is visible for the species *Bromus inermis, Agrostis alba* and *Phleum pratense* belonging to *accumulators* (non-hyperaccumulators). Though in these species metals did not concentrate in the shoots we can't ignore the fact of metals accumulating in great amounts in the roots of these plants. As it is known from literature data, phytoimmobilization is a containment technology that aims at immobilizing contaminants in the root zone. In phytoimmobilization it is also important to reduce the translocation of pollutants to the shoots to avoid further transfer into the food-chain [3]. *Dactylis glomerata, Bromus inermis, Agropyron repens, Agrostis alba* and *Phleum pratense* accumulated Zn and Pb in most amounts in the strong contaminated soils mainly in the roots. Therefore these species can be the candidates for using them for *phytoimmobilization.*

Table 1. Pb influence on growth and biomass 21-days seedlings of grass species

Plant species, Pb, mg/kg	Length of shoots		Dry weight of shoots		Length of roots		Dry weight of roots	
	cm	%	mg	%	cm	%	mg	%
1	2	3	4	5	6	7	8	9
Setaria viridis								
Control	5,18±0,02	100	1,33±0,004	100	2,94±0,08	100	0,34±0,02	100
450	4,46±0,23	86	1,0±0,02	75	2,13±0,08	72	0,28±0,01	82
900	4,32±0,09	83	1,00 ±0,03	75	1,87±0,06	64	0,26±0,01	76
Agrostis alba								
Control	3,8±0,002	100	0,095±0,0001	100	0,50±0,01	100	0,25±0,001	100
450	2,70±0,02	71	0,095±0,0001	100	0,4±0,01	80	0,27±0,01	108
900	2,60±0,02	68	0,075 ±0,0001	79	0,37±0,01	74	0,25±0,01	100
Agropyron repens								
Control	12,40±0,15	100	1,25±0,021	100	6,30±0,02	100	0,36±0,002	100
450	10,4±0,1	83	1,16±0,008	93	4,4±0,03	70	0,28±0,004	78
900	9,10±0,08	73	1,15 ±0,01	92	3,7±0,01	59	0,28±0,003	78
Bromus inermis								
Control	11,98±0,49	100	29,10±0,11	100	12,05±0,19	100	13,93±0,16	100
450	11,63±0,20	97	22,6±0,08	78	8,04±0,01	67	8,36±0,13	60
900	10,33±0,16	86	20,04 ±0,01	69	5,35±0,16	44	7,06±0,13	51
Phleum pratense								
Control	4,56±0,09	100	4,30±0,11	100	1,41±0,11	100	2,36±0,09	100
450	3,43±0,10	75	3,58±0,01	83	1,37±0,01	97	1,79±0,01	76
900	2,9±0,09	64	2,60±0,05	60	1,14±0,04	81	0,79±0,01	33

2. Effects of Pb and Zn on Growth and Biomass of Grass Species in Hydroponic Conditions.

It was studied the influence of lead (450, 900 mg/l) and zinc (350-700 mg/kg) on growth and biomass of grass species in hydroponic conditions.

The shoot growth of *Bromus inermis* and *Setaria viridis* was retarded at 900 mg Pb/l by 14 and 17% only. Shoot growth *Phleum pratense* at the same concentration of lead was inhibited by 37% (Table 1).

We observed that at the most Pb concentration in the species which had a strong retard of the shoot growth the suppression of root growth was less. The length of roots at this concentration of Pb decreased in the almost reverse order: *Phleum pratense> Agrostis alba> Setaria viridis> Agropyron repens> Bromus inermis.* Interestingly, that the species which have the most level of retardation of shoot growth (*Phleum pratense*) at the same time have the lowest level of suppression of root growth, and vice versa (*Bromus inermis*). The shorter the shoots the longer the roots. Probably, it is the adaptation response of plants to the abundance of metals in the medium.

Table 2. Zn influence on growth and biomass 21-days seedlings of grass species

Plant species, Zn, mg/kg	Length of shoots		Dry weight of shoots		Length of roots		Dry weight of roots	
	cm	%	mg	%	cm	%	mg	%
1	2	3	4	5	6	7	8	9
Setaria viridis								
Control	4,3±0,09	100	0,86±0,003	100	2,6±0,01	100	0,22±0,009	100
350	4,2±0,11	98	0,86±0,02	100	2,5±0,13	96	0,18±0,08	82
700	4,0±0,07	93	0,84 ±0,03	98	2,4±0,09	92	0,18±0,001	82
Agrostis alba								
Control	3,72±0,002	100	0,11±0,005	100	0,40±0,008	100	0,04±0,001	100
350	2,90±0,03	78	0,1±0,001	91	0,36±0,01	80	0,02±0,003	50
700	2,20±0,02	59	0,09 ±0,0003	82	0,30±0,01	75	0,01±0,01	25
Agropyron repens								
Control								
350	17,10±0,09	100	1,7±0,0005	100	7,50±0,005	100	0,56±0,005	100
700	10,1±0,06	59	1,6±0,003	94	3,89±0,07	52	0,28±0,004	50
	8,96±0,07	53	1,45 ±0,01	85	3,1±0,01	41	0,26±0,003	46
Bromus inermis								
Control	11,98±0,49	100	29,10±0,11	100	12,05±0,19	100	13,93±0,16	100
450	9,51±0,04	79	21,00±0,20	72	7,08±0,24	58	6,23±0,03	45
900	9,89±0,35	83	17,50 ±0,09	60	7,09±0,09	59	5,20±0,11	37
Phleum pratense								
Control	4,56±0,09	100	4,30±0,11	100	1,41±0,11	100	2,36±0,09	100
450	3,37±0,11	74	2,93±0,04	68	1,28±0,01	91	1,58±0,01	67
900	2,99±0,05	66	2,06±0,07	48	1,18±0,02	84	0,80±0,01	34

The biomass of shoots under Pb stress decreased in the order: *Agropyron repens> Agrostis alba> Setaria viridis>Bromus inermis> Phleum pratense;* of roots - *Agrostis alba > Agropyron repens ≥ Setaria viridis > Bromus inermis> Phleum pratense.*

The shoot growth of *Bromus inermis* and *Setaria viridis* was retarded at 700 mg Zn/l by 7 and 17%, accordingly. The length of *Agropyron repens*'s roots and shoots at the same concentration of Zn was lowest relatively control plants. The growth of roots of *Setaria viridis* was inhibited only by 8% (Table 2).

The tolerance to Zn on the basis of shoot biomass under Zn stress decreased in the following order: *Setaria viridis> Agropyron repens ≥ Agrostis alba > Bromus inermis> Phleum pratense*. Biomass of *Setaria viridis*'s roots decreased in the lowest rate, of *Agrostis alba* – in the most rate.

Thus, *Phleum pratense* was most sensitive to Pb and Zn. *Agropyron repens* was most tolerant to Pb. *Setaria viridis, Agropyron repens* were most tolerant to Zn. Other species have an intermediate position on the tolerance.

3. Lead and Zinc Content in the Plant Parts of Grass Species

The most accumulation of lead at 900 mg/l was observed for the roots of *Ph. pratense* (90 281 mg/kg), *Ag. repens* (77 135,6 mg/kg) and *Br. inermis* (69 991,8 mg/kg). The lowest amount of lead was found in the roots of *S. viridis* (33 974,8 mg/kg). The most amount of Pb accumulated in the shoots of *A. alba* (22 670 mg/kg), low amount - in *Ag. repens* (2 091,3 mg/kg).

The most content of lead was found in the roots. The shoot/root ratio of all species was <1 and decreased in the order: *Agrostis alba* (0,47) >*Setaria viridis*(0,17) > *Phleum pratense* (0,06) >*Bromus inermis* (0,04)>*Agropyron repens* (0,027).

Thus, it was revealed that *Phleum pratense, Agropyron repens* and *Bromus inermis* accumulate great amounts of Pb in the roots in comparison with other species.

The most amount of Zn at 700 mg Zn/l was in shoots and roots of *Br. inermis* and *Ph. pratense*. The lowest amount - in *Ag. repens* and *A. alba*. Concentrations of Zn in the shoots decreased in the order: *Bromus inermis > Phleum pratense >Setaria viridis> Agrostis alba >Agropyron repens*; in the roots: *Phleum pratense >Bromus inermis >Setaria viridis≥ Agropyron repens >Agrostis alba.*

The study of zinc accumulation by grass species showed that all species except *Agropyron repens* accumulated Zn mainly in the shoots at 700 mg Zn/l. Shoot/root ratio of Zn content of all species except *A. alba* and *Ag. repens* was >1 and decreased in the order: *Bromus inermis* (1,8) > *Setaria viridis*(1,3) > *Phleum pratense* (1,14)> *Agrostis alba* (0,92) > *Agropyron repens* (0,46).

A. alba accumulated Zn approximately in equal amounts in the roots and shoots; *Ag. repens* - mainly in the roots.

Phleum pratense and *Bromus inermis* distinguished by high accumulation of Zn in the shoots and roots.

Thus, in hydroponic conditions at its high concentrations Pb was accumulated by species mainly in the roots and shoot/root ratio was < 1. At high concentration of Zn it was accumulated mainly in the shoots except *Agropyron repens. Agrostis alba* accumulated Zn approximately in equal amounts in the roots and shoots.

Phleum pratense, Bromus inermis distinguished by high accumulation of Zn in their parts. These species accumulated most amounts of Pb in the roots. The consequence of high accumulation of HM for these species was a low tolerance to these metals. On the contrary, *Agropyron repens* was more tolerant to Zn and Pb, however it accumulated the great amounts of Pb in the roots. *Setaria viridis* was the most tolerant to Zn, however it accumulated the significant amount of Zn in its parts.

The screening of grass species of Kazakhstan flora on heavy metal accumulation ability have shown the high accumulation capacity of plant roots. The metal concentrations in the shoots and roots of *Dactylis glomerata* proportionally increased with the increasing metal content in the soils. In our opinion *Dactylis glomerata* belong to *indicators*.

Metal excluders effectively prevent metals penetration into aerial parts over a broad range of metal concentrations in the soil; however, they can still contain large amounts of metal in their roots. Therefore *Agropyron repens* can be preliminarily attributed to *excluders*. As excluders do not in general regulate metal uptake, restriction of transport from roots to shoot is the likely mechanism of control [9].

We have not found the species convenient for p*hytoextraction* process. But the results obtained allow to hope that such species as *Dactylis glomerata, Bromus inermis, Agrostis alba* and *Phleum pratense* can be used for *phytoimmobilization.*

Agropyron repens can be used as a cattle forage because of low accumulation of heavy metals in the shoots.

REFERENCES:

[1] M.S Panin. Influence of antropogenic activity and human argochemical activity on migration of heavy metals in system "soil-plant": *Int. Conf. "The state and rational use of soils in Kazakhstan"*. Almaty, 1998. P. 76-79.

[2] Chaney R.L., M. Malik, Y.M.Li, S.L.Brown, J.S Angle, A.J.Baker: *Curr. Opinions Biotechnol.*. 8, 279 (1997).

[3] W.Wenzel, D.Adriano, D.Salt, R.Smith: Phytoremediation: Bioremediation of Contaminated Soils, *Agronomy Monograph* no 37. Madison, USA. 1999. 508 p.

[4] D.E Salt, M. Blaylock, P.B. Kumar, V. Dushenkov, B.D. Ensley., I. Chet and I. Raskin: *Biotechno*l. 13, 460(1995).

[5] A.J Baker: *J. of Plant Nutr.,* 3 (1-4), 643(1981).

[6] R.R Brooks (Ed.): Wallingford. UK: CAB International. 1998.

[7] S.P. McGrath: In: Brooks R.R. ed., Wallingford., UK:CAB International, 261 (1998).

[8] A.J.Baker, P.L. Walker (Ed.): Boca Raton, CRC Press, 155 (1990).

[9] W. Ernst: *Proc.Int.Conf. on Heavy Metals in the environment.* Toronto, Canada. 11, 121(1975).

In: New Research on the Environment and Biotechnology ISBN 1-60021-042-2
Editors: A. Kuznetsov and G. Zaikov, pp. 17-27

Chapter 2

BIOTECHNOLOGY INTERNET MARKETING FOR PROVISION OF NATURE RESOURCES RATIONAL USAGE AND ENVIRONMENTAL SAFETY

E. A. Demcheva[1], E. I. Khabarova and A. V. Demchev
M.V.Lomonosov Moscow State Academy of Fine Chemical Technology
119571 Moscow, Prospect Vernadskogo 86

ABSTRACT

The internet market of biotechnology is one of the most ambiguous branches. The development of this internet branch is not constant and varies depending on each direction. Concerning the biotechnological branch as a whole, it cannot be characterized as advanced; however, some directions are allocated from the general picture and possess their own unique solutions and technologies. One such branch segment is the market of environmental biotechnology.

Keywords: biotechnology, environment, environment problem solution, information resources, internet-marketing

INTRODUCTION

Every manufacturer constantly has the task of satisfying not only existing consumers, but also the future inquiries of consumers. Regular correlation of the company activity purposes, which taking into account constantly changing political-economical conditions of the country and also of the world, means not only management perfection, but also continuous, purposeful re-structuring and technical modernization of manufacturing.

Certainly, it is necessary to take into account that new technologies are characterized, as a rule, by superior characteristics. However, in the beginning, this advantage can be opposed by

[1] E-mail: khei-mitxt@mtu-net.ru, Fax: (095) 434-87-11

huge expenses and a deficiency of know-how. There is also a high probability that their introduction will cause resistance both in the company and at the market [1].

How should it be interpreted in the context of the last decades?

First of all, it is necessary to recall that in the last quarter of the last century plenty of information demanding its distribution and practical usage has appeared.

Secondly, it is impossible to disregard the fact of an ecological imperative development concerning the same period (basic concept of ecological ethics), according to which the vigorous people's activity of uncontrollable and unguided technological civilization development, a set of restrictions has been imposed (technological, economic, social), which humanity should not to break in connection with possible catastrophic consequences both for biosphere and for Homo sapiens.

Thirdly, it is relevant to notice that practically at the same time the population of many advanced countries began to show serious concern in unsatisfactory quality of the life, conformity of an inhabitancy to the needs of particular individuals, social layers, countries and regions ceased to be provided.

Due to such course of events, vector of scientific and technological progress development have changed its direction and a social landscape of a human life.

In the beginning of the 70's of last century an information-technological revolution has come in a peaceful way. Cyberspace began to be formed almost all over the world. Computers, communication systems, genetic decoding and programming – have all served for amplification and expansion of human ideas [2].

For the last ten years of the "biotechnological boom" the quantity of environmentally friendly technologies, simulating natural processes and using for this purpose inherent natural ecosystem components or their life cycle products, including the obtainment of target products, transformation of the waste products, clearing of the polluted natural resources, restoration of the broken grounds, etc. has been sharply increased. [3]

At the same time an essential new strategy for advancing industrial enterprises has been developed. This strategy is about the creation of environmentally friendly manufacturing (synonyms: "cleaner production", "ecologically sound manufacture"), meaning the continuous application of preventative environmental protection strategies in the technological processes and production with the purpose of reducing the production's negative influence on people's health and on the environment [4]. During the introduction of this strategy there is a change from the principle "polluter is paying" to the principle "prevention of pollution is profitable" [5].

All the factors above have served as a prologue to the formation of ecologically oriented biotechnological developments on the Russian internet-market.

PRELIMINARY REMARKS

The internet market of biotechnology can be related to one of the most ambiguous branches. The development degree of this internet branch is not constant and varies depending on each direction. Concerning the biotechnology branch as a whole, it cannot be characterized as advanced, however, some directions are allocated from the general picture and possess their own unique solutions and technologies. One of such branch segment is the market of environmental biotechnology, considered further in more detail.

Preliminary research has been conducted in 2002. Its purpose was revealing a saturation of the Internet information resources with the biotechnology oriented content and in particular offers of environmental biotechnology i.e. marketing monitoring of the Russian market in the field of biotechnology development applied for the solution of environmental problems. During that research more than 3,500 web-sites from which more than 35 % contained the necessary information have been analyzed. Such a high parameter has allowed the conclusion about expediency of such research in 2003-2004 for revealing dynamics of internet market development in this field. Upon conducting this research it was found that actually there are no similar offers in the market of environmental biotechnology. At the same time, demand for such works is rather great and is mainly satisfied with an interested company's own forces. Thus, it was possible to discover a huge market branch where this research is the only choice nowadays.

The Scheme of Research

Internet monitoring was carried cut in three stages:

- The first wave of monitoring has been conducted from February, 1 until April, 1 2002 (duration - 2 months)
- The second wave of monitoring has been conducted from February, 1 until April, 1 2003 (duration - 2 months)
- The third wave - from March, 1 until May, 1 2004 (duration - 2 months)

Monitoring was carried out with the help of such internet search engines as Yandex, Rambler, Google, and Aport. Forty five keywords have been selected as the objects of search, concerning the subject. An additional condition of search was viewing at least the first 70-100 references of each search engine's responses. For the convenience of further analysis all found information has been classified in 9 basic sections: biotechnology dictionaries, departmental biotechnology information portals, branch biotechnology information portals, thematic web-sites, biotechnology conferences, biotechnology magazines, foundations – grant distributors, biotechnology fairs and exhibitions, ecology-actual biotechnology developments.

The significant volume of the information was collected during this monitoring, allowing us to analyze the development of biotechnology information resources and environmental biotechnology offers in the internet for several years (since February 2002 till May 2004). Thus, the results of this monitoring are of a great value.

Information Resources

The informational saturation of this branch in the internet can be defined as satisfactory. Large information projects and balanced solutions - presence of the information on the market and branch analytics are available. Alongside with such structure, catalogues and databases on the companies and scientific laboratories / institutions of the current branch are widely used. It is necessary to pay attention as to an integration of information and trading solutions, such as the number of resources which are positioned, as trading, containing the wide list of information materials and can be used, not only for contact with the possible partners, but also for reception of the information.

Upon the investigated period the significant internet-market space growth of biotechnology-oriented information resources and the biotechnology developments focused on the solution of environmental problems were observed. Results of monitoring have shown, that web- conferences (a number for two years has increased more than 7-fold), web-dictionaries (a number for two years has increased 5,5-fold) and web-sites of magazines (for two years their number has increased 5-fold) are considered to be the most developing category (Fig. 1). On the contrary revealing any significant changes among web-sites of foundations – grant distributors and web-sites of fairs and exhibitions on biotechnology was not possible. Their number in the internet upon the investigated period has changed slightly.

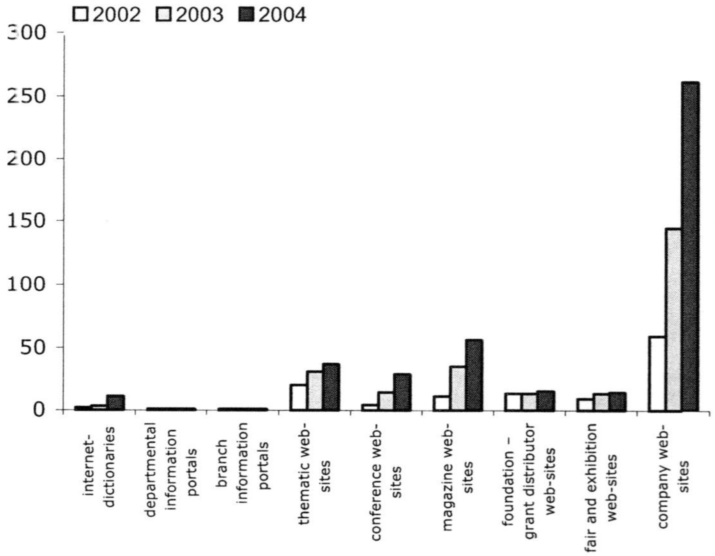

Figure 1. Dynamics of environmental oriented information recourses in the field of biotechnology on the internet

Also significant changes have occurred in the category of the companies and scientific laboratories / institutions which offer biotechnology developments for the solution of environmental problems.

In the market investigated the main share belongs to the web-sites of the companies and scientific laboratories which offer biotechnology development for the solution of environmental problems - their share in the whole structure has made 61,4 % by 2004. Among all other categories some group of research participants are noticeable by the amount of web-sites: thematic web-sites (13,2 %), web-sites of magazines (8,7 %), and also web-sites of conferences (6,8 %), but in comparison with 2002 their share in the main structure has decreased (Fig. 2).

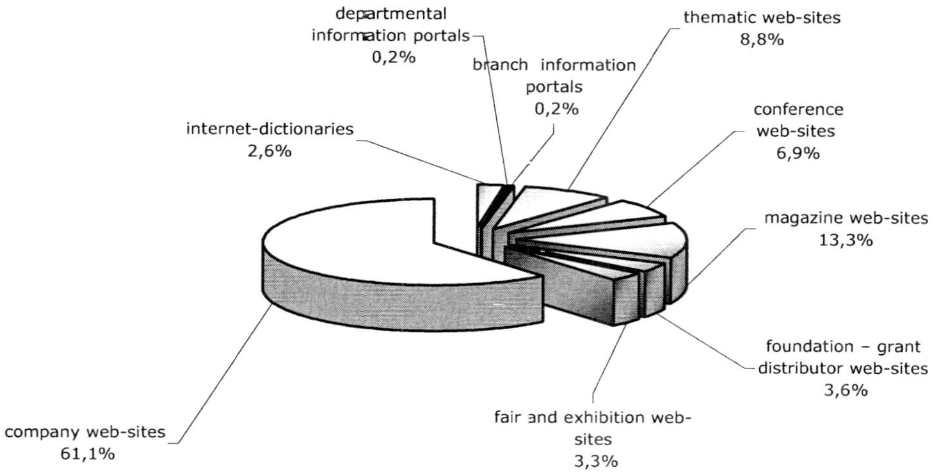

Figure 2. Structure of environmental oriented information recourses in the field of biotechnology on the internet in 2004

Except for those specializing in providing information about the market, it is possible to note the resource catalogues of the biotechnology branch. Branch and departmental information portals in the main structure are the minority.

Biotechnology Oriented Information Solutions

With examples of information resources in the network it is possible to name a number of projects:

- Internet-dictionaries of biotechnology
- Russian biotechnology web-portal
- Web-catalogue - Scientific and Technical Information Network
- Constantly updated web-servers delivering the different information for the biotechnology branch
- Web-sites, giving the operative information on biotechnology conferences
- Web-sites duplicating printed production
- Web-sites, informing on biotechnology grants
- Web-sites of exhibitions and the fairs propagandizing achievements of biotechnology

At the same moment all these solutions have the news and survey (analytical) information, data on branch events and actions. Information solutions of the branch are not just limited to this list. However, it would be desirable to pay attention to some of them.

ENVIRONMENTAL BIOTECHNOLOGIES

This direction develops in the Russian internet pretty fast. Probably, it is so because of the active growth and development of this branch in Russia as a whole. Marketing research of the Russian internet-market of biotechnology developments, focused on the environmental problems solution, has reflected their quantitative growth from 74 up to 336 positions (more than 4-fold) for two years (from February 2002 till May 2004). The tendency of the further growth of the offers in this market in 2005, according to the results at the current moment shows that in 2005 the number of web-sites with similar offers will continue to increase (Fig. 3).

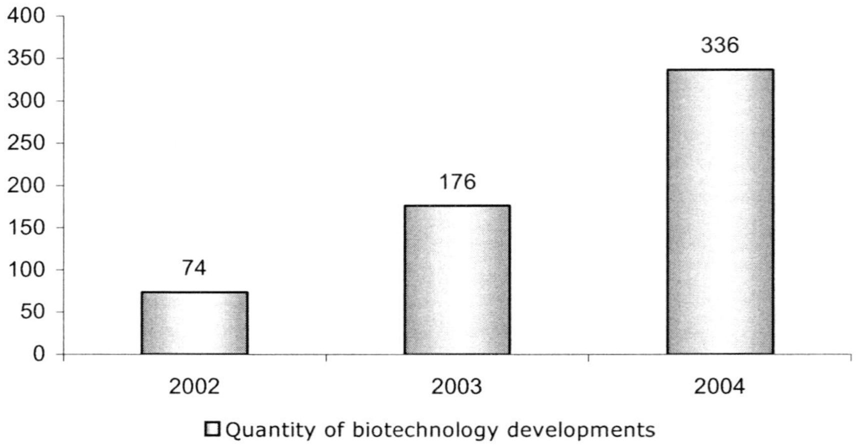

Figure 3. Dynamics of biotechnology developments in the internet

Various directions of the environmental biotechnology market have their own offers in the network and common for all of them are the universal information projects which contain materials in various directions of the market and are suitable for both, experts of the market, and simple visitors. A number of them combine a service and information directions. Some biotechnology oriented companies use the internet for organizing the sales process and communication with their partners. This can be a proof of the interest of such companies to the internet.

The analysis of 336 biotechnology developments for the solution of the environmental problems presented in internet for the period since February 2002 till May 2004, has shown, that the amount of the approaches using potential of biology objects for decreasing and liquidation of negative consequences of anthropogenic activity on the basis of the traditional and special equipment, makes up already a significant figure (62 %) (Tab. 1). Hence, it is possible to ascertain the fact of gradual formation of the Russian internet-market of the biotechnology developments focused on the solution of environmental problems that mankind is facing nowadays.

Table 1. Ways of environmental problem solution.

Ways of an environmental problem solution	**Number of projects in 2002**	**Number of projects in 2003**	**Number of projects in 2004 ***
Increasing in extraction of mineral resources	2	5	7
Processing of organic waste into biogas	7	13	22
Processing of organic waste into liquid fuel	1	1	1
Processing of organic waste into protein concentrate	4	4	7
Processing of organic waste into nutrients and veterinary medicines	1	1	1
Processing of organic waste into bio-fertilizers	3	12	28
Transformation of organic waste into compost	7	18	33
Biochemical clean up of soil	10	24	33
Biological re-cultivation of polluted sites	3	7	15
Microbiological purification of air from gas emissions	2	4	11
Biochemical sewer waters treatment	23	50	97
Bio-monitoring	11	24	45
Increase of stability and increase of productivity of agricultural crops	0	5	11
Struggle against bio-damages	0	7	24
Liquidation of solid household waste	0	1	1
Total:	**74**	**176**	**336**

* Data gathering on 2004 has been conducted during the grant "EcoLogic technologies" organized by CAF Russia by initiation and financial support of BAT Russia.

The received statistical results once again confirm the process of modern business philosophy development: to create something new, necessary for the consumer. This innovative philosophy reflects new public needs and the behavior of the market, inclined to:

- the active type of adaptation to new managing conditions
- continuous search of innovations and manufacture diversification
- technological and social and economic innovation
- the active involvement in this process of the physical and intellectual resources

There are 119 web-sites of biotechnology developers that have been revealed all over the internet by May 2004. It is about 4-fold more than those in 2002 (Fig. 4). Web-sites both of the companies and scientific laboratories/institutions are presented among them. Developers create their own web-sites to use them as an interactive information channel of marketing for creation of a positive image for their company, interaction with surroundings and promotion of the goods and services.

The analysis of the web-site's information content and structure of the Russian companies offering biotechnology developments and production for the environmental problems solution have been conducted by a technique, allowing the definition of web-activity factors of the branch and the separate enterprises, and also corresponding factors in special directions of information content of the company's web-representations [6].

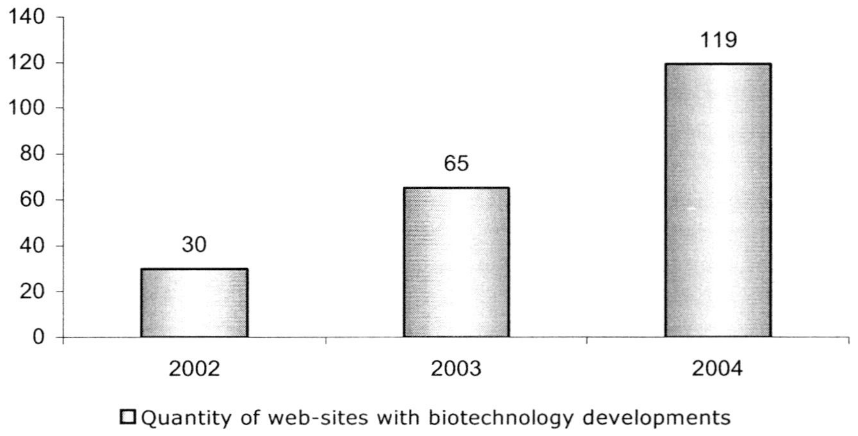

Figure 4. Dynamics of web-sites with biotechnology developments on the internet

The analysis has shown that the majority of these company web-sites were loaded considerably fast, but they often lack convenient navigation and functionality of the menu. Thus, it has appeared that:

- sitemap is organized for 29 % of all web-sites;
- searching engine is organized for 20 % of web-sites;
- forum or the guest book is organized for 17 % of web-sites;

For all company web-sites there is a standard structure of navigation through the main menu.

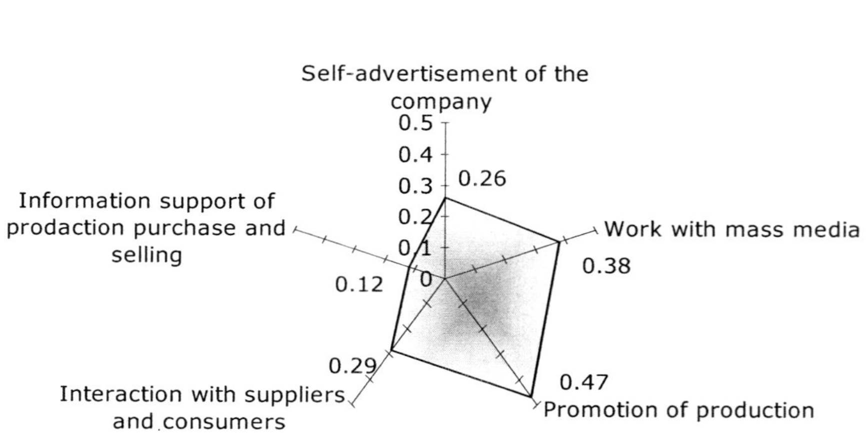

Figure 5. Web activity of biotechnology companies in the internet

Basically, the companies are using the internet for presentation purposes, i.e. for representation of the information on their activity, history of the company, plans of the further development and the re-structuring, production, coordinates and the contact information. Also this analysis has shown that companies pay the most attention to the general information about company activity in their web-sites (Fig. 5). Interaction with the mass media is developed rather weakly on those web-sites. The most attention is focused on product promotion. Interaction with suppliers and consumers on such web-sites are developed rather weakly as well. Informational support on purchases and sales are developed in the weakest way.

Newly formed macro-economical tendencies are leading to major changes in modern business, mainly in the style of company communications with clients and society. Also such tendencies will appear very soon in the more effective way of using the internet advantages by the companies, and declaring principles of ecological management on the pages of their web-sites.

Conducted analysis of companies specializing in the field of biotechnology developments for environment protection has shown that companies of this branch are not using the whole advantages of the internet for the development of new forms of business interaction. Nowadays the web-sites of such companies are mainly oriented on PR direction i.e. are mainly used for general presentations and are not a serious tool for business dealing. A newly formed tendency of internet usage for interactive communication with suppliers and consumers appears insufficient. This allows concluding that Russian companies of this branch are not ready for dealing over the internet.

It is possible to assume that unavailability of dealing over the internet by the Russian companies specializing in the field of biotechnology developments for environment protection is caused by such factors as: the personnel qualification, problems with internet access, the unavailability of consumers and suppliers to the new forms of interaction, lack of laws regulating e-business, low needs for usage of new trade channels, presence of commercial information which cannot be freely distributed, and the safety problems of data transmission via internet.

But the first step is already made. Presence of the web-site itself is a kind of interest in innovative business methods. Most resources are directed for presentation of general information about company, as well as some for marketing and sales development. This can possibly be interpreted as a fact that biotechnology companies' interest for the internet, as for the new trade channel, will soon gain more precise outlines. And the strategy of e-business development will become a constituent of the company's main strategy.

Besides creation of their own web-sites, many biotechnology developers use specially created solutions. Environmental biotechnology cannot be referred to the leading branches by the amount of such solutions. However, by now it is possible to find a number of projects which allow the companies to interact via internet. Almost 40 % of the biotechnology developers who did not have their own web-site are using such solutions and placing there the information concerning their developments.

It is necessary to note the additional services which are provided to clients of such solutions. First of all, there is the information supply. Another convenient service is the client support service which allows people to ask questions about the web-page function and to receive quality assistance from the online specialist.

The overwhelming majority of biotechnology developers offer the one-two environmental biotechnology solutions, but also there are such developers who provide 6 or more solutions.

Conclusions

The branch of environmental biotechnology is one of the most quickly developing branches of the internet because of such facts as:

- a number of general information and special solutions;
- plenty of the companies are working using the internet;
- wide spectrum of information-analytical materials of the branch.

While the majority of other branches are represented on the internet by Moscow companies, in this particularized market there are plenty of regional companies.

Concerning demand for special projects, it is possible to admit that for the companies' manufacturers and sellers this is one of the most effective solutions allowing to advertise and sell their production through the internet - the majority of such companies are registered on several web-sites.

Many companies have not created their own internet strategy and therefore there are a lot of difficulties in this way of interaction. One of the weakest critical moments is the unsuccessful organization of interaction with potential clients and visitors, but there are a the number of solutions which provides optimal mechanisms for information gathering and reception of feedback. Thus the future changes of internet activities of biotechnology companies can be predicted.

In summary it is necessary to pay attention to that fact that unfortunately, today the companies of many branches try to improve their competitive positions by carrying out the strengthened energies in the field of development of base technologies or overdo research in the field of key technologies while new biotechnologies, which in the near future can become progressive or the most important, are frequently neglected at the process of planning.

References

[1] D. I. G. Shneider. *Technological marketing*. Yanus-K, Moscow, 2003. 478 p. (in Russian)

[2] M. Kastelie. *Information epoch: economy, society and culture*: Translate from English, under redaction of O. I. Shkaratan. GU VSHE, Moscow, 2000. 608 p.

[3] T. E. Popova, E. V. Popova. *Biotechnology and society*. Science, Moscow, 2000. 108 p (in Russian)

[4] S. I. Pronin, L. E. Taratinova. Ecological and economic transformation of the industrial enterprises on the basis of pure manufacture strategy: *Safety of life*, 5(2002). p. 4-10.

[5] M. Gluhova. Calculation of financial effect at introduction resourse-saving and nature protection actions in energy: *Practical marketing,* 7 (77), 2003. p. 12-16.
[6] *www.e-commerce.ru/reviews/metall/web_repr.html*

In: New Research on the Environment and Biotechnology ISBN 1-60021-042-2
Editors: A. Kuznetsov and G. Zaikov, pp. 29-36

Chapter 3

The Bioremediation of the Chemically Contaminated Ecosystems

E. I. Ermakov*[1]*, G. G. Panova and O. A. Stepanova
Agrophysical Research Institute, Russian Academy of Agricultural Sciences
Grazhdansky prospect, 14, 195220, Saint Petersburg, Russia,

Abstract

The strategy of the bioremediation of ecosystems is offered for chemically polluted zones by a highly toxic propellant – asymmetrical dimethylhydrazine. It is grounded on the results of complex research of interaction's regularities of soils, plants and concomitant microorganisms with the toxicant in the sites of fallen rocket parts and on the specially built polygons in natural and regulated conditions.

Keywords: propellant, asymmetrical dimethylhydrazine, soils, plants, microorganisms, bioremediation.

In the modern conditions of global anthropogenous pollution of the natural environment a priority task is the development of both effective and ecologically safe methods of detoxication and restoration of chemically polluted territories based on the knowledge about laws of the interaction between ecosystems with their elements and pollutants.

The importance of contamination problem is getting more obvious when highly toxic and reactive xenobiotics penetrate into natural habitats, such as ingredients of a fluid propellant on the basis of asymmetrical dimethylhydrazine (ADMH).

Pollution of territories by ADMH has both local (the locus up to 100 meters in sites of its accidental spilling out and the falling of separating rocket stages), and extensive character (many tens kilometers at atmogenic, and also edaphic migration with entry into ground waters, the rivers and other water reservoirs) [1].

[1] raes@agrophys.ru, fax +7(812)5341900

In polluted ecosystems the plants absorb ADMH and products of its destruction from soil and directly from the atmosphere. The degree of fixation of propellant components in soil - plant complex is high enough, and xenobiotics are involved in natural cycle of substance in agro- and ecosystems. Oxidation and destruction of ADMH is accompanied by formation of rather toxic compounds, among them such as nitrosodimethylamine (NDMA), dimethylamine, tetramethyltetrasen, dimethylmethylenhydrazine, formaldehyde, and also amines, ammonia, ethane, ethylene, hydrogen, nitrogen and other ones [2-4]. The majority of them have cancerogenic and mutagen properties and they are dangerous to the health of people and animals [5,6].

Consequences of pollution of ecosystems by ingredients of rocket fuel up till now were estimated according to the contents of toxicants in natural environments. Methods of bioindication practically were not applied, and for this reason influence of ADMH and products of its destruction on live substance and their transmission on trophic chains were poorly investigated. There is no information about the role of plants and microorganisms in detoxication of soils polluted by ADMH.

Methods of elimination of chemically polluted soils offered to the present time are rather various. Among them biological methods are ecologically most harmonic, that is why they have an obvious advantage as compared to physical, chemical and physicochemical methods [7,8]. However, bioremediation, particularly bound with the introduction of a selected microorganism's strains into soil and biologically active substances, often does not produce wishful results. It is difficult to realize, especially in conditions of the North region [9-11 and others].

The purpose of the presented work is to research the regularities of influence of ADMH and products of destruction on plants, microbial community in natural and regulated conditions, and also to find out the capabilities of complex living organisms in the process of utilization of these toxicants. The strategy and methods of bioremediation of chemically contaminated ecosystems were worked out on the basis of the received knowledge.

Material and Methods

Processes of interaction of soil - plant systems with ingredients of rocket fuel were investigated - in sites of falling stages of rockets on the trasses of their flight (the Arkhangelsk region) and on base of expressly built polygons on a space port "Plesetsk" and in laboratory of plant's light physiology and biophysics in Agrophysical Research Institute. In sites of falling stages of rockets we have estimated visually a vegetable cover and a mesofauna. We have studied the reaction to contamination of the microorganisms living on thallus of lichens, and also soil microorganisms in the peatbogs.

As subjects of investigation on polygons we used soils, plants, and microbal communities. As root-inhabited medium we used a model organo-mineral analog of soil created on the basis of high-moor peat of a low degree of decomposition and mineral components [12-14], and also natural soils – soddy podzolic sandy loam from Leningrad region and podsolic alluvial-humus from the Arkhangelsk region. On polygons in field and regulated conditions we cultivated barley plants of the varieties Belogorsky and Pirkka, and pea plants of the varieties Fitotron 1 and Nemchinovsky 96. The number of barley plants of each varieties per variant was 40 on laboratory polygon and 400 on polygon of a space port "Plesetsk", pea plants – 30 and 200 accordingly. On polygon of a space port "Plesetsk", the plants were grown in special vegetation installations with the natural illumination and temperature under a light-transparent polymer roof open from all sides. On this polygon, besides researches of regularities of penetration of ADMH and products of its destruction into the economic useful part of plants at rather high pollution of soils, we found out capabilities to bring down the content of xenobiotics in plants by not root processing them using the solution of the siliceous chelated microfertilizers (SCM) created in Agrophysical Research Institute [15].

At cultivation of plants in laboratory of Agrophysical Research Institute on the basis of a regulated agroecosystem [16] temperature of air and soil was kept at 20-25°C, the light period was 14 h a day, the intensity of the radiant flux in range of PAR was 100±10 $W \cdot m^{-2}$, the relative humidity of the air was 60-70 %, the soil moisture – 60 % of the maximum moisture capacity. During one year in such conditions we received three - four crops of plants that has allowed to accelerate at least in 3 times the study of the transformation of ADMH and products of his destruction into soil - plant system as compared to natural conditions. The doses of the pollution of soils by ADMH corresponded to the ones at the places of falling of the separated parts of rockets and were 10; 25; 50 and 100 g ADMH per kg of soil on the polygon of a space port "Plesetsk" and 1.0; 2.5; and 5.0 on the laboratory polygon. Uncontaminated soil was used as a control. The replication of trials in the experiments was 3-5 times.

During a vegetation period we made phenological observations of plant's development, and we have estimated main parameters of their growth and productivity. The quantification of microorganisms of the different taxonomical and physiological groups we determined by a method of seeding on solid and liquid mediums conventional in microbiology [17, 18]. The quantity of the bacteria using organic forms of nitrogen was determined on meat peptone agar (MPA); using preferentially mineral forms of nitrogen - on starch-ammoniacal agar (SAA); oligonitrophylic - on mediuim of Eshbi. The calculation of actinomyces was done using SAA. The microfungi were selected on agar beer-wort and on the medium of Chapek (pH 4-5). Total number of bacteria was determined using the method of a direct counting on membrane ultrafilters (Sinpor 8) after they coloured by erythrosine (x1300). Furthermore we differentiated bacteria by morphological composition, we have measured average volumes of bacterial cells and estimated cell biomass of bacteria per unit of soil mass. All data on the quantity of the microorganisms were calculated as an average value of minimum of three repetitions.

The contents of ADMH and its derivates in soils and plants were determined according to the methods developed at the Institute of Biophysics of the Russian Ministry of Health and at the Russian Scientific Center of Applied Chemistry [19].

RESULTS AND DISCUSSION

The results of visual observations of plants and mesofauna, researches of plants and microorganisms in sites of falling the stages of rockets made it possible to conclude that in most cases there are no negative consequences of ecosystem's pollution by rocket fuel's ingredients on the biota. The reaction of microorganisms both in peats, and on the surface of the lichen's thallus taking into account heterogeneity of a microrelief and a composition of the vegetation was practically the same and was determined basically by ADMH and NDMA content during three years after local pollution of ecosystems by rocket fuel ingredients. We have found the tendency of stimulation of the biological processes based on the data on the quantity and multiplicity of microorganisms (see table). In microorganism's community the bacteria dominated by the quantity, and at the increasing of ADMH content in soil the trend of increasing the bacterial cell's total number, cell biomass of bacteria was observed.

The dominating forms of bacterial community in the control were rod-shaped and coccus-like forms – 65-73 % and 27-35 % respectively. At ADMH content within the range of 0.9-37.7 mg / kg of peat a morphological diversity of bacterial cells was more obvious than in the control. The presence of rod-shaped cells of the bigger size was observed. In samples of peat with maximal content of ADMH the share of rod-shaped forms of cells was 94 %, coccus-like – 6 %.

The quantity of microfungi in sites where ADMH was spilled was bigger than in uncontaminated ecosystems, except for the variant with the maximal content of ADMH in the peat substrate where the activity of fungi was suppressed completely. Apparently, first of all it is stipulated by inhibiting influence on them of products of ADMH destruction - formaldehyde and ammonia, and also with the increase of pH of the peat substrate up to 9.5 as compared to 5.0 in the control. It is known, that the best values of pH for the majority of fungi lays below 7. On the surface of lichen's thallus in places where the rocket fuel was spilled the quantity of fungi decreased during the time. Judging by the quantity of microfungi in ecosystems with older term of pollution by ADMH, it is possible to make a conclusion about ability of natural ecosystems in conditions of the North region fast enough to indemnify for the damages caused by the presence of rocket fuel components.

The analysis of the group composition of microorganisms in peats and on a surface of lichen's thallus in contaminated natural ecosystems revealed a preferential increase of the number of the microorganisms using easily accessible forms of the organic and mineral nitrogen, as contrasted to the observed ones in the control, that indicates the enrichment of substrates by these substances as a consequence of the rocket fuel spillage. The change of the number of microorganisms of the mentioned groups in peat positively correlates with amount of ADMH in it ($r=0.92$, $n=3$, $P=0,95$). The trend of displacement of the group of the oligonitrophilic bacteria that are characteristic for the deficiency of the nitrogen substrates was noticed.

In sites of falling the stages of rockets, on polygon of the space port "Plesetsk" and in laboratory conditions (a regulated agroecosystem) the similarity of biota reaction to peats and soils pollution by ADMG and products of its destruction was detected. The experimental data obtained by us have allowed to range the toxicant's concentration in the ecosystems. Taking into consideration the consequences of the influence on peas and barley, on the microbal community in soils and on plant's roots, the amounts of ADMH per 1 kg of soil can be conditionally classified as low ones – less than 0.1 g, moderate – up to 10 g, high – from 10 up to 50 g, and very high – above 50 g. Respectively, effects are observed ranging from stimulation of the growth and productivity of plants and the activity of concomitant microorganisms to suppression of their vital activities and even lethality.

Table. Microorganisms in peat and on a surface of thallus of lichens (*Cladina cetraria*) in sites of rocket's stages

Time after polluting the area by ADMG, months	The content, mg / kg of dry peat		Total number of bacteria, milliard cells/g dry mass	Volume of bacterial cells, mkm^3	Biomass of bacteria, mg/kg of dry mass	Quantity of microorganisms, x10^6 / g of dry mass			
	ADMG	NDMA				The bacteria using organic forms of nitrogen	The bacteria using mineral forms of nitrogen	Oligonitrophilic bacteria	Microfungi
Peat									
Control	0	0	6.50±2.40	0.069±0.001	0.449±0.009	0.81±0.07	4.04±0.410	0.17±0.010	0.03±0.005
1.2	6763.6	6.2	58.40±3.40	0.071±0.001	4.146±0.127	10.73±0.87	62.20±2.800	0.12±0.020	0
12	37.7	0.8	98.40±18.30	0.067±0.001	6.593±0.980	5.04±0.50	29.52±2.520	0.19±0.020	0.24±0.040
24	4.4	0.3	42.60±10.90	0.068±0.001	2.897±0.600	1.76±0.11	9.40±0.700	0.07±0.020	0.49±0.080
36	0.9	0.1	15.90±0.70	0.066±0.001	1.049±0.045	2.88±0.27	5.18±0.720	0.13±0.020	0.22±0.030
LSD_{05}			9.02	0.002	0.693	0.70	2.32	0.02	0.06
Lichen's thallus									
Control	-	-	0.09±0.01	0.080±0.009	0.007±0.001	1.22±0.22	0.16±0.020	0.16±0.030	0.02±0.010
1.2	-	-	0.13±0.01	0.169±0.006	0.021±0.002	3.76±0.26	0.01±0.003	0.01±0.002	0.09±0.020
24	-	-	0.10±0.01	0.082±0.006	0.008±0.001	1.25±0.20	0.14±0.020	0.11±0.020	0.05±0.010
LSD_{05}			0.03	0.009	0.002	0.30	0.02	0.03	0.02

Thus, with the content of ADMH in the amount below 0.1 g per kg of soddy-podzolic sandy loam and model organomineral soils, the accumulation of the overground matter in barley plants of the varieties Pirkka and Belogorsky increased by 151 and 138%, respectively, of the grain mass from the plant — by 162 and 229%, while in pea plants of the varieties Fitotron 1 and Nemchinovsky 96 by 43 and 33%, and in the economically useful part by 127 and 44%. At the content of the ADMH up to 10 g per kg of model organomineral soil we observed the decrease of the mass of the grain in barley of the Pirkka variety for 66% and of the Belogorsky one for 70%, in the pea plants of the variety Fitotron for 29%, and of the variety Nemchinovsky 96 for 37%. In the range of ADMH concentrations in this soil from 10 to 50 g per kg we noted retarded development of the plants, and the decrease of the growth and productivity parameters. At the content of ADMH in the amount above 50 g per kg of soil the plants perished.

A growth in the contamination of the soils by NDMH was attended by an increase and prevalence of the fraction of bacteria using readily accessible organic and mineral forms of nitrogen in their metabolism, that corresponded to the generally known scenario of the behavior of microbe communities when an additional readily accessible organic substrate enters into the soil. The denitrification processes ran more actively, autotrophic nitrification and the activities of the microorganisms less adapted to the changing conditions of the environment (microfungi, actinomycetes, aerobic cellulose-decomposing bacteria) were suppressed. With time, obviously, as the content of the available substrate diminished, the number of microorganisms using readily accessible forms of nitrogen and carbon decreased, while the number of microorganisms capable of mineralizing high-molecular organic compounds increased. All these processes had fluctuating features.

We have established a close relation between the nature of the change of the biological activity of contaminated soils relative to the control and the intensity of the destruction of the toxic substance in the course of the experiment (the correlation coefficient is 0.95-0.99 for the soddy-podzolic sandy loam soil and 0.60-0.87 for the model organomineral soil). Thus, in the period of intense destruction of the ADMH in the contaminated soil we observed considerable activation of the activity of the microorganisms of the dominating groups, in the period of stabilization of the content of the toxic substance in it, the number of most groups of the microorganisms did not differ reliably from that one in the control.

The study on the dynamics of ADMH content in model organomineral soil and in plant's roots during nine vegetation periods has shown, that toxicant destruction was maximal at the initial stages after pollution of soil, then it slowed down essentially. At the end of the observation the amount of xenobiotic in soil under barley was on a level, close to the maximum concentration limit (according to the standards for soils: 0.1 mg of ADMH/kg), in soil under peas it exceeded the maximum concentration limit 8-fold, and in soil without plants - more than 20-fold.

Intensity of ADMH and its derivates destruction in the root-inhabited environment at the favourable light and temperature conditions at growth and development of barley and peas plants increased by 2-10-fold (differences are reliable at 5% level of confidence, n=3). It is fixed, that such effect is connected with the plant's ability to absorb and transform ADMH, to involve the resulting products of destruction into the plant's metabolism, to stimulate the activity of accompanying microorganisms, both directly and indirectly to intensify the processes of transformation of organic compounds in the soil.

Thus, there are all reasons to consider the higher plants together with a concomitant biota as an efficient biogenous natural factor in a solution of the problem of the bioremediation of chemically contaminated territories. Apparently, for more efficient phytoremediation of contaminated soils it is perspective to use plants taken from native flora, fitting in the best way to habitat factors and tolerable to the toxicant's influence.

During the researches on polygon of the space port "Plesetsk" the capability to decrease the content of ADMH and products of its destruction in plants was shown. It was reached by means of nonroot treatment of the plants at definite stages of their ontogenesis with solutions of the siliceous chelated microfertilizers created in Agrophysical Research Institute [20]. Even at enough high level of soil's pollution, the content of ADMH and products of its destruction in economically useful part of barley and pea's crop decreased by 2-3-fold (differences are reliable on 5 % a level of confidence, n=3).

Knowledge received as a result of complex researches has formed the basis for the development of the strategy of bioremediation of chemically contaminated ecosystems. It rests on V.I.Vernadsky's doctrine [21] about living material being a permanently functioning chemical force on surfaces of our planet and takes into account the gained experience of restoring phytomelioration of degraded soils in agro- and ecosystems [22, 23].

At forming the bioremediation strategy of the degraded ecosystems we took into account the ability of ADMH to the intensive evaporation and to have toxic influence on warm-blooded organisms via a gas phase. To restore the lands in places of technogenic catastrophes it is offered to place a calculated layer of the organomineral model soil with a high specific surface area reaching 300-400 $m^{2} \cdot g^{-}$ [14] onto a surface of soil contaminated by ADMH. On the top of this layer special containers with seeds of plants are placed. In process of plant's growth the roots with a concomitant biota penetrate into the ADMH contaminated layers of soils and intensify toxicant destruction, and the products formed are involved in metabolism. Subsequently cultivated plants will be substituted by the representatives of indigenous flora possessing, as a rule, higher adaptable surroundings-forming ability [12, 13].

The important element of the developed strategy of the bioremediation of chemically contaminated ecosystems is the decrease of the content of xenobiotics in economically useful part of cultural and wild plants in sites of falling the stages of rockets and in zones with allocation of specific dangerous chemical factories, e.g. using new compositions of SCM. The purpose of the use of them is to reduce more efficiently the penetration of ADMH, products of its destruction or other toxicants into plants, and also to compensate in zones of the ecological risk both deficiency of microelements and the quality of vegetative production [15].

In summary it is necessary to remark, that the developed strategy and methods of efficient bioremediation of ecosystems in places polluted by ADMH and by other toxicants have doubtlessly an advantage from both ecological and technological points of view and it can guarantee the high environment protection effectiveness because it provides for the use of the natural biodestructors – higher plants with concomitant microorganisms. It is a good perspective to solve complex problems of environment protection, to rise the stability of agrarian manufacture, to recover population's environment and to improve the quality of life in ecological risk areas.

REFERENCES

[1] N.S. Kasimov, V.B. Grebenjuk, T.B. Koroleva, Y.V. Proskurjakov: *Pochvovedenie*. 9. 110 (1994). (in Russian).

[2] S. Sarner: *Chemistry of rocket fuel*. Mir, Moscow, 1969. 488 p. (in Russian).

[3] B.V. Ioffe., M.A. Smiths, A.A. Potekhin: *Chemistry*, Leningrad, 1979. 223 p. (in Russian)

[4] P.E. Tulupov, S.V. Kolesnikov, V.P. Kirjukhin: *Proceedings of IV All-Union meeting «Pollution of atmosphere and soils»*. Hydrometeoizdat, Moscow, 1991. 87. (in Russian).

[5] Monograph on the *evaluation of carcinogenic risk of chemicals to man* / Agency for Research on Cancer. Lion, 4. 137 (1974).

[6] K.C. Back, V.L. Cowter, A. Thomas: *Aviation space and environmental medicine*, 49 (4), 591 (1978).

[7] J. Blackman, C. William: *Basic hazardous waste management*. Boca Raton, Florida, 1992. 339 p.

[8] *The Superfund innovative technology evolution program*: Technology Prfiles. Tifth Edition. Washington, 1992. 388 p.

[9] D.G. Zvyagintsev (Ed.): *The microorganisms and the preservation of soils*. Publishing house of the Moscow State University, Moscow, 1989. 206 p. (in Russian)

[10] T.V. Koronelli: *Applied biochemistry and microbiology*, 32 (6), 579 (1996). (in Russia).

[11] G. Stotzky, H. Babich: *Sci. Reports*, 7 (4), 163 (1986).

[12] E.I. Ermakov, A.E. Ermakov, Yu..I. Zheltov, etc.: RF Patent № 2073417, *Byull. Izobret*. 5, 1997. (in Russian).

[13] E.I. Ermakov, A.E. Ermakov, Yu..I. Zheltov, etc.: RF Patent № 2101898, *Byull. Izobret*. 2, 1998. (in Russian).

[14] E.I. Ermakov, G.G. Panova: *Russian Agricultural Sciences*, 10, 20 (2000). (in Russian).

[15] E.I. Ermakov, S.H. Sembaev: *The Agrarian Science*,1, 18 (2002). (in Russian).

[16] E.I. Ermakov, in: *Productional process of plants in regulated conditions*. Hydrometeoizdat, St. Petersburg, 3, 1993. (in Russian).

[17] N.S. Egorov (Ed.): *Practical work on microbiology*: Studies. Publishing house of the Moscow State University, Moscow, 1976. 307 p. (in Russian).

[18] *Methods of experimental mycology*: Naukova dumka, Kiev, 1982. 550 p. (in Russian).

[19] V.K. Potrohov, V.G. Pimkin: *Proceedings of scien.-and-pract. conf. «Ecological aspects of influence of ingredients fluid missile fuel on the environment»*. Russian Scientific Center Applied Chemistry, St. Petersburg, 1996. 72 p. (in Russian).

[20] E.I. Ermakov, I.V. Medvedeva: USSR Inv. Cert. № 743641, *Byull. Izobret*. 24, 1980. (in Russian).

[21] V.I. Vernadsky: Nauka, Moscow, 1965. 374 p. (in Russian).

[22] Z.Sh. Shamsutdinov: *Kormoproizvodstvo*, 4-6, 17 (1993). (in Russian).

[23] B. Ensley: *Mat. Agr. Biotechnol*. Concil, Ithaca (NY). 8, 145 (1996).

In: New Research on the Environment and Biotechnology
Editors: A. Kuznetsov and G. Zaikov, pp. 37-49
ISBN 1-60021-042-2

Chapter 4

STIMULATION OF THE ACTIVITY OF OIL OXIDIZING MICROORGANISMS IN CONTAMINATED SOIL

Ekaterina Pleshakova[1], Ekaterina Dubrovskaya and Olga Turkovskaya

Institute of Biochemistry and Physiology of Plants and Microorganisms, RAS
13 Pr. Entuziastov, 410049 Saratov, Russia

Mineral fertilizers azofoska, molasses, and a synthetic surfactant (SS) were used to stimulate the development and activity of hydrocarbon-oxidizing microorganisms (HOMs) in soils with different ages and types of pollution. It was showed that the application of molasses, both singly and in combination with other supplements had the most effect on the increase in the activity of the oil-oxidizing microflora in the freshly polluted soil. In the case of the soil with the 10 years age residual-oil pollution, the application of supplements that facilitate dispersion and availability of a pollutant was needed to enhance of HOMs.

Key words: oil-polluted soil, bioremediation, hydrocarbon-oxidizing microorganisms, stimulants

INTRODUCTION

In recent years, researchers have increasingly been looking at *in situ* biostimulation as their method of choice for the cleanup of oil-polluted soils. An important step in the use of this method is the choice of expedients for the intensification of the functional activity of natural microbial cenoses, especially the group of hydrocarbon-oxidizing microorganisms (HOMs).

Several works have been published that address the development of methods for stimulating natural biodegradation, including the choice of optimum conditions for the mineralization of oil hydrocarbons, in particular by improvement of aeration, polluted-soil

[1] Telephone 845-2-97-04-03. Fax 845-2-97-03-83,E-mail: ecbio@ibppm.sgu.ru

humidity, temperature and nitrogen-nutrition conditions, application of emulsifiers, and so on (Wang et al., 1990; Freijer, 1996; Freijer et al., 1996; Singh and Stapleton, 2002).

Many investigators have reported on a positive effect obtained upon addition of supplements, mostly nitrogen and phosphorus (Kireyeva, 1994; Margesin et al., 2000; Kao et al., 2001; Rahman et al., 2003). Others noted no enhancement of cleanup; on the contrary, they found that the addition of supplements has a negative effect (Margesin et al., 1999).

The application of organic material has a positive effect on the degradation of oil hydrocarbons, increases the number of the bacterial population, and enhances microbial activity (Yermolenko et al., 1997; Jorgensen et al, 2000; Vasudevan et al., 2001).

Certain companies produce patented biological supplements that, according to their advertisers, are stimulatory to a broad range of microorganisms capable of utilizing many pollutants.

Yet, most investigations note that the estimation of the implications of ecosystem pollution by oil and the direction of the way to their liquidation should be done with account taken of the level of pollution, the composition of oil, the age of the pollution, the soil properties, and the landscape and climatic conditions (Glazovskaya, 1988). Each specific case should have its own combination of expedients in order to achieve the best result.

In connection with the aforesaid, the aim of this work was to find optimal supplements to increase the numbers and activity of HOMs in soils with different ages and types of pollution.

MATERIALS AND METHODS

A bench-scale experiment was conducted that used microcosms with two soil samples: one freshly polluted by crude oil (concentration, 50 g/kg) and the other with an "old" (~10 years) residual-oil pollution (initial concentration of oil products, ~380 g/kg).

The characterization of the fractional composition of the oil pollution of the original soil samples is shown in Table 1.

Some agrochemical and biological characteristics of the soils studied are listed in Table 2.

The soil (1,5 kg) was placed in plastic containers. The cleanup was performed at room temperature (19-24^{0}C), with occasional loosening and moistening of the soil and with addition of various stimulants. To stimulate the development of HOMs, we used the most available and industrially cost-effective supplements: the mineral fertilizer azofoska, which is a nitrogen and phosphorus source; molasses as an additional, readily available source of carbon and biologically active substances; and a synthetic surfactant (SS) as an emulsifier. The substances were applied singly and in combinations (Table 3).

The same supplements were used for the soil with an old residual-soil contamination. Alongside the entire set of supplements, one of the samples also received crude oil at a concentration of 0.5 g/kg.

The total numbers of heterotrophic microorganisms, ammonifiers, nitrifiers and dinitrifiers were determined by conventional methods using the appropriate nutrient media (Yegorov, 1976).

Table 1. The ratio between oil fractions in the original soil samples (%)

Fractions	Oil pollution	Residual-oil pollution
Paraffins and naphthenes	68.2	75.4
Mono- and bicyclic aromatic compounds	11.8	7.4
PAHs	7.5	9.0
Alcohol-benzene resins	12.6	8.2

Table 2. Agrochemical and biological characteristics of the soils studied

Parameters:		Clean soil	Soil with an old residual-oil pollution
Content of substances:	Total C, (%)	1.39	23.90
	Water-soluble C (% of total C)	1.08	0.18
	Nitrate N (mg/kg)	6.0	0
	Ammonium N (mg/kg)	12.0	15.0
	Mobile P_2O_5 (mg/100g)	20.0	9.7
Microbial numbers, cell/g of soil:	Total number of saprophyte microorganisms	1.6×10^6	2.0×10^5
	HOMs	6.5×10^5	4.0×10^4
	Ammonifiers	2.5×10^6	2.9×10^5
	Nitrifiers	1.7×10^4	3.5×10^3
	Dinitrifiers	2.5×10^5	5.2×10^4
	Phosphate-solubilizing bacteria	2.0×10^4	1.0×10^3
pH		7.28	7.54

Table 3. Treatment in the remediation of oil-polluted soil

Treatments	**Supplements applied (g/kg of soil)**
1	control (no supplements)
2	azofoska (0.7)
3	molasses (10.0)
4	SS (0.02)
5	azofoska (0.7), molasses (10.0)
6	azofoska (0.7), SS (0.02)
7	molasses (10.0), SS (0.02)
8	azofoska (0.7), molasses (10.0) and SS (0.02)

The hydrocarbon-oxidizing bacteria were found by the membrane-filter method (Chekalov, 1991). The phosphate-solubilizing bacteria were counted by the Muromtsev (Muromtsev, 1957) modification of Gerretsen´s method (Gerretsen, 1948).

The soil content of total and soluble carbon was determined as described in (Malakhov, 1984). The content of the mobile phosphorus P_2O_5 in mg/100g was monitored by a method using a hydrochloric-acid soil extract (Malakhov, 1984). The nitrate-N concentration in the soil microcosms was determined as described in (Yagodin et al., 1987). The nitrite-N concentration was measured as described in (Andreuk et al., 1988).

The oil-product fraction was analyzed by the Polunina and Kushina method (Ryl'kov, 1977).

Results and Discussion

Most studies conducted in various bioclimatic zones suggest that the soil microorganisms response to oil pollution is an increase in their gross numbers and an enhancement of their activity. The increases in the numbers and activity first occur in microorganisms carrying out the initial metabolism of hydrocarbons. HOMs are nonspecific and can assimilate a wide range of organic compounds (carbohydrates, proteins, or lipids); that is, they are euthytrophic (Koronelle, 1996). Because of this, microorganisms belonging to this group occur widely and are a normal component of an unpolluted system as well. Specifically, in the clean soil used by us in our biostimulation experiments, the initial HOM content was 6.5×10^5 cells/g.

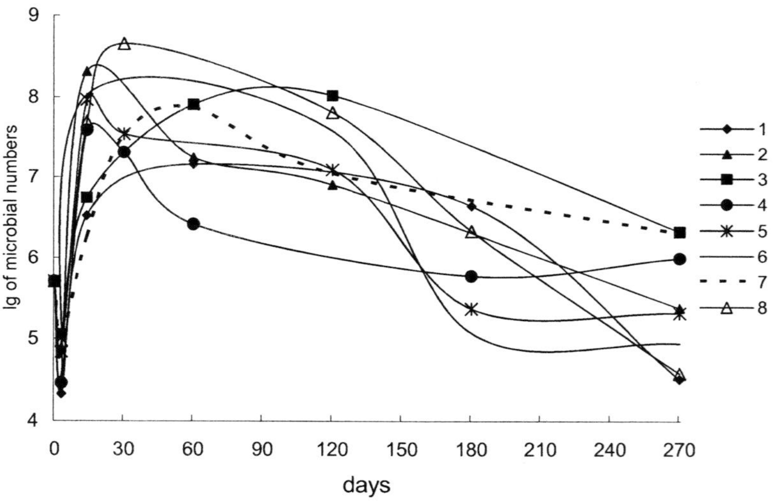

Fig. 1. Time course of the HOM numbers in the oil-polluted soil

Analysis of our data on changes in the numbers of oil-oxidizing microorganisms in various treatments in the experiment using a fresh oil pollution (Fig. 1) showed that at 3 days after addition of oil and stimulants and after the beginning of agrotechnical measures, the HOM numbers decreased, sometimes by an order of magnitude.

Most probably, this effect resulted from the toxic action of oil on the microorganisms. The maximum numbers of HOMs were observed on 14-30 day of the experiment in most of the soil model objects. In treatments with oil-polluted soil (2, 3, 5-8), the microbial numbers in this group reached values around 10^8 cells/g of soil. The greatest increase in the numbers of HOMs during their activization was observed in Treatment 3 upon addition of molasses (from 6.5×10^5 to 1.0×10^8 cells/g of soil during 4 months of cultivation) and in Treatment 8 upon addition of molasses simultaneously with the SS and mineral fertilizer (an increase to as high as 4.5×10^8 cells/g during 1 month). In control Treatment 1 (no supplements), the maximal increase in the HOM numbers was 22-fold in comparison with the initial content, whereas in Treatment 8 (several supplements) the microbial numbers increased 696-fold. This finding allowed us to conclude that the application of supplements does more to enhance the development of oil-oxidizing microorganisms than do agrotechnical measures. By 6 months of biostimulation, the HOM numbers decreased to the initial values in some cases and remained stable, with some fluctuations, until the end of the experiment. In Treatment 4 (with the SS as the sole supplement), the development of the oil-oxidizing microorganisms was minimal throughout the experiment, in comparison with the other treatments.

The initial numbers of HOM in the soil with the old residual-oil pollution were lower (4.0×10^4 cells/g) than those in the clean soil. In most of the samples of this soil, as distinct from the freshly polluted samples, the maximum increase in the HOM numbers was observed at later than 30 days of cultivation (Fig. 2). This increase (~ by 2 orders of magnitude) was not so noticeable as it was in the freshly polluted soil, and no considerable decrease in the HOM numbers occurred throughout the experiment. We proposed that the long period of the pollution contributed to the selection and adaptation of HOMs and to the stabilization of the microbial community. Therefore, the supplements caused not so much an increase in the HOM numbers as an enhancement of microbial metabolic capabilities, which was subsequently verified in estimating the decrease in the soil content of the oil products.

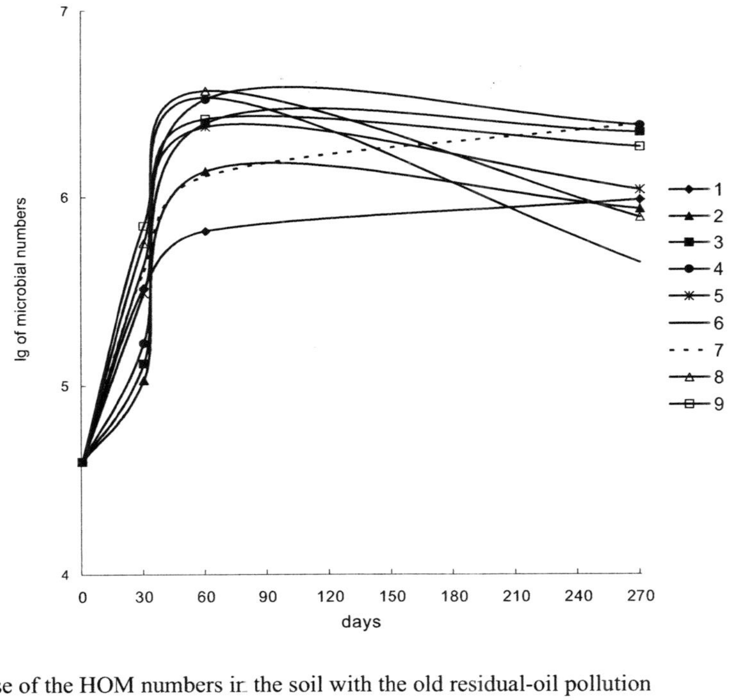

Fig.2. Time course of the HOM numbers in the soil with the old residual-oil pollution

In Treatments 4, 6, and 8, the HOM numbers increased the largest after 2 months of cultivation (83.5, 85.2, and 92.7 fold, respectively). The data for Treatments 4 and 6, in which the SS was applied singly and in combination with the mineral fertilizer, were opposite to the results of the experiment on freshly polluted soil, which were characterized by animal stimulatory effect of the SS on the increase in the HOM numbers. The development of HOMs was least noticeable in Control Treatment 1 (soil with the old residual-oil pollution).

Determination of the total number of saprophytic microorganisms (TNSM) is an important criterion for the assessment of the state of the processes in soil. Any effect on soil causes a change in the soil microbiota. In this work, the application of oil was also stimulatory to the development of the saprophytic microflora as a whole. The addition of molasses, both singly and in combination with the other stimulants, caused the greatest increase in the numbers of the soil microbial community and in the numbers of the HOMs. In all treatments, the highest bacterial numbers were fond over a period of 14-30 days. The addition of the entire complex of supplements to the polluted soil increased the TNSM from 1.615×10^6 to 3.6×10^9 cells/g of soil. The increase was the smallest in the control treatment without addition of the stimulants (to as high as 4.6×10^7 cells/g of soil after 30 days of cultivation). The application of the SS even somewhat suppressed the development of the saprophytic microflora.

The TNSM in the soil with the old residual-oil pollution increased from 2.0×10^5 cells/g to $4.3\text{-}5.0\times10^8$ cells/g in most of the treatments including supplements, after 30-60 days of cultivation. The TNSM´s are very similar in different treatments, and it is hard to single out the most stimulatory supplements. Only in control Treatment 1 (no stimulants) the development of saprophytic microorganisms was less pronounced in comparison with the other treatments. It is noteworthy their similarly to the development of HOMs in this soil, the activization of the saprophytic microflora was stable throughout the experiment, with no appreciable decrease in its numbers.

Thus, the additions undoubtedly enhance the development of HOMs and other heterotrophic microorganisms in the soil with the old (10 year) residual-oil pollution. However, the best results were achieved when the SS was applied singly and in combination with the mineral fertilizer. For the freshly oil polluted soil, the best results came from the addition of molasses singly and in combination with the other supplements. In either case of pollution, Treatment 8 (with all the supplements) provided optimum conditions for the development of HOMs and other saprophytic microorganisms.

Analysis of the decrease in oil products in the microcosms studied yielded interesting results. For the freshly polluted soil, the maximum decrease in oil content was achieved upon application of all the supplements in Treatment 8 (38.8% for 6 months of cultivation and 46% for 9 months cultivation) (Fig. 3). After 9 months of recultivation, close values were found in treatments including molasses used singly and in combination with mineral elements; and a little lower values were found in the treatment "SS + mineral fertilizer". In these samples, a correlation between the increase in the HOMs numbers and the decrease in oil content was most noticeable. The application of the SS had a minimal effect; during 9 months of the experiment, the decrease in the content of oil products was only 23.7%, almost equal to the control value (23%). By 9 months, the degree of oil biodegradation in these treatment has increased, but still it was the lowest in comparison with the other treatments (36% in the supplement-free treatment and 37% upon addition of the SS as the sole component).

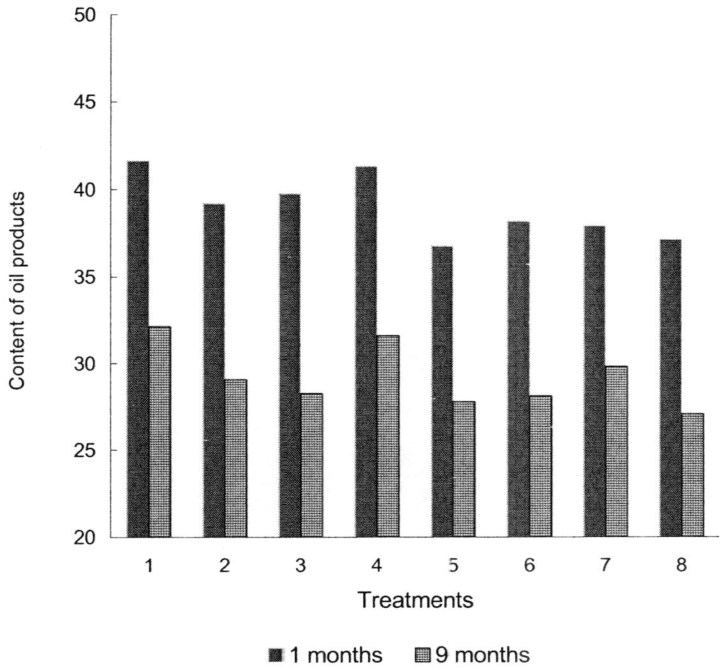

Fig 3. Results of the enhancement of oil degradation (initial concentration, 50 g/kg) in the soil. Treatments: 1, agrotechnics; 2, agrotechnics + mineral fertilizer; 3, agrotechnics + molasses; 4, agrotechnics + SS; 5, agrotechnics + fertilizer + molasses; 6, agrotechnics + fertilizer + SS; 7, agrotechnics + molasses + SS; 8, agrotechnics + molasses + fertilizer + SS

Analysis of the content of individual fraction in residual oil products after 9 months of the experiment showed that in all the treatments, the lowest decrease was that in the content of mono- and bicyclic aromatic compounds: from 62% (control, no supplements) to 70.4% (treatment with compounds) (Fig. 4). The degradation of the paraffin-naphthene fraction ranged from 26.7% (control) to 42.8% (molasses + fertilizer). PAHs were more resistant to biodegradation.

The Maximum degradation (by 27.3%) obtained upon application of the fertilizer and the SS; the minimum degradation (17.3%), upon application of molasses and the SS. Alcohol-benzene resins were degraded even worse: by 4.1% (control) and by 26.7% (fertilizer + molasses). Overall, the following tendency was noted: upon addition of molasses (an additional source of carbon and biologically active substances), the available oil fractions – mono- and bicyclic aromatic compounds and paraffin-naphthene hydrocarbons – were degraded to a greatest extent; upon addition of the mineral fertilizer we observed better degradation of PAHs and alcohol-benzene resins.

In the case of the old residual-oil pollution, we obtained different results. The maximum decrease in the content of oil products was noted upon addition of the SS as the sole supplement and also upon addition of crude oil alongside the entire complex of stimulants (62.2 and 61.7% for 9 months, respectively) (Fig. 5).

Additionally, we did not find any effect of supplement addition on the preferential degradation of this or other fractions; the level of utilization of all the fractions was the same (Fig. 4). The supplement combinations that gave the best results for the fresh oil pollution yielded even a negative results in the case of the old residual-oil pollution. Upon addition of molasses, the decrease was 37%; upon simultaneous addition of molasses, fertilizer, and the SS, 26.8%; and in the supplement-free control, 40.4%.

Possibly, the presence of the SS and crude oil facilitates the emulsification and dissolution of heavy oil products, making them more available to microorganisms.

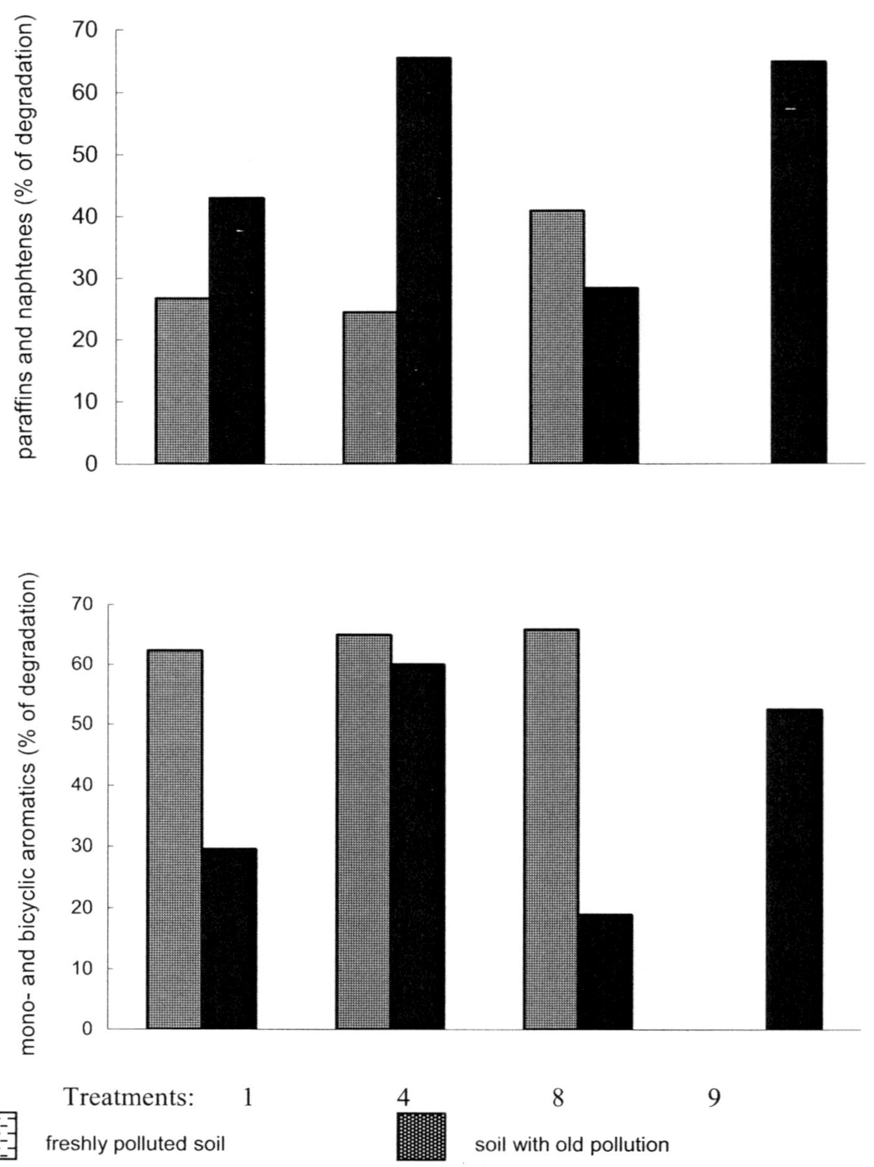

Fig. 4. Continued on next page

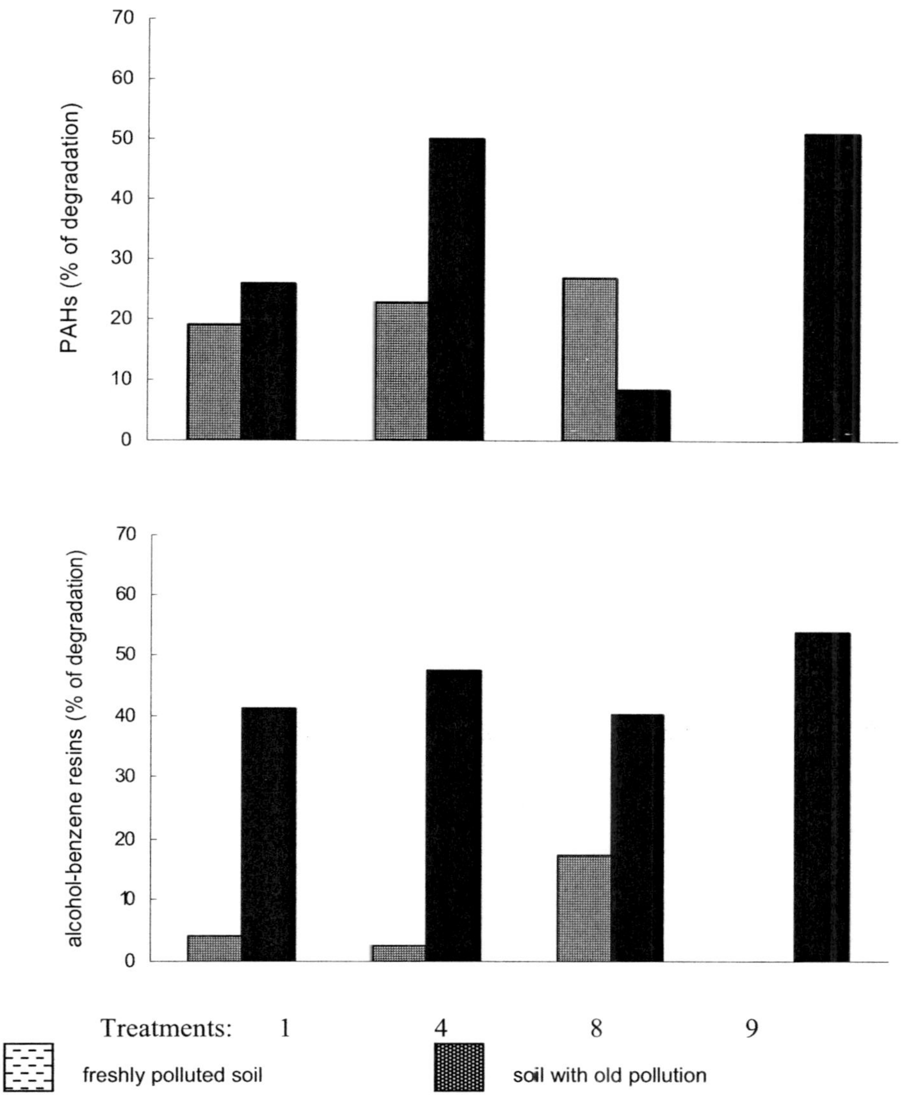

Fig. 4. Degradation of oil fraction in the soil (fresh oil pollution and old residual-oil pollution) after 9 months of remediation (%).

The complete mineralization of oil hydrocarbon can be characterized, to some extent, by a decrease in total and soluble carbon. We showed that a correlation existed between the decrease in the oil quantity and smooth decrease in total and soluble carbon in the freshly oil-polluted soil. At the highest decrease in treatments using molasses (more than 50%) it is observed the decrease by 45% for total carbon and 22% for soluble carbon, due to accelerated development of microflora in this soil and, hence, increased consumption of carbon reserves.

Conversely, in the case of the old residual-oil pollution, no noticeable changes in total carbon content occurred in any of the treatments, whereas after 9 months the concentration of soluble carbon increased. The highest increase was noted in treatments using the SS, singly and in combination with mineral elements (2.4-2.7-fold). The tendencies found can be explained by the oxidation of oil hydrocarbons in the soil, resulting in the formation of a noticeable quantity of water-soluble components.

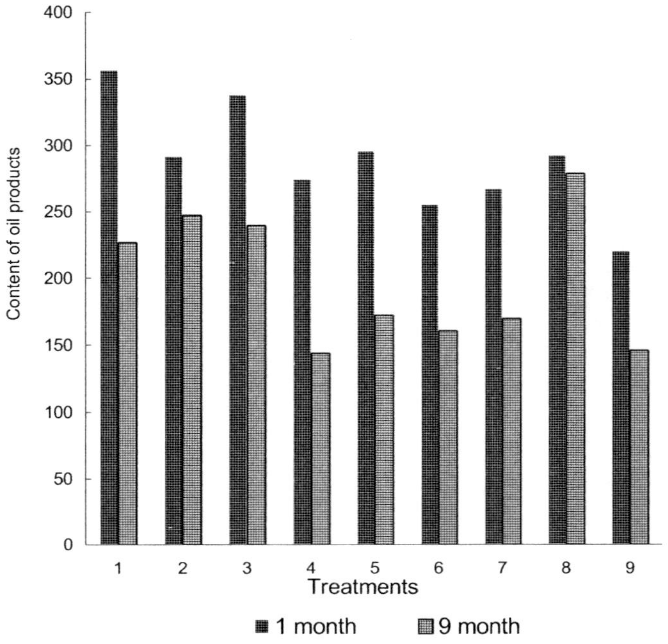

Fig 5. Results of the enhancement of oil-product destruction in the soil with the old residual-oil pollution (380 g/kg). Treatments: 1, agrotechnics; 2, agrotechnics + mineral fertilizer; 3, agrotechnics + molasses; 4, agrotechnics + SS; 5, agrotechnics + fertilizer + molasses; 6, agrotechnics + fertilizer + SS; 7, agrotechnics + molasses + SS; 8, agrotechnics + molasses + fertilizer + SS; and 9, agrotechnics + fertilizer + molasses + SS + crude oil

When a soil is polluted by oil, it receives a lot of carbon, and the C/N ratio changes sharply. To ensure that there is a physiological balance, which is essential for normal bacterial growth, we applied to the soil a mineral fertilizer as a nitrogen, which is the most available to most microorganisms, was utilized almost completely during 2 weeks of the experiment. The ammonium-nitrogen reserves were maintained probably at the cost of transformation of fixed forms to free forms. The ammonium-nitrogen content was 10-50 mg/kg of soil up to 6 months of recultivation. It is known that the application of additional mineral-nutrition elements is particularly important early in biorecultivation (Kireyeva, 1994); then, the microbial community can function at the expense of the processes occurring therein. We also found that a repeated application of mineral supplements did not increase the HOM number or the degree of oil degradation in polluted soil, in comparison with the control samples, when did not received any additional supplements.

In the case of residual-oil pollution, the ammonium-nitrogen content did not decrease throughout the experiment, because of the lower rate of expansion of the microbial community.

The results of determination of the mobile-phosphorus content in the soil showed that the P-content were not depleted during 9 months of cultivation and were 42-28 mg/100 g of soil in different treatments in the freshly polluted soil and 11.75-25.0 mg/100 g of soil in the soil with the old residual-oil pollution.

It should be noted that the phosphate-solubilizing microorganisms, whose numbers increased noticeable after 2 weeks of recultivation, transform low-soluble P-compounds into soluble forms, thus facilitating the supply of the bacterial community with available P-forms. On the other hand, the period of active growth and development of all members of a microbial community is attended by a considerable consumption of mobile phosphates. In this connection, it proved hard enough to find a relationship between the results obtained on the soil content of mobile phosphorus and the numbers of phosphate-solubilizing microorganisms in the soil and the total number of saprophytic microorganisms. By 9 months of cultivation, the number of phosphate-solubilizing microorganisms returned to the original values, and the mobile-P content was higher than that in the original soil. This finding allowed us to assume that additional mobile-phosphate reserves came from the transformation of various P-forms occurring in the soil.

CONCLUSION

In summary, the supplements used by us for soil remediation had different effects on the increase in the number and activity of the oil-oxidizing microflora, depending on the type and age of the oil pollution. For the enhancement of the development and activity of HOMs in the freshly polluted soil, the optimal effect was achieved by the application of molasses, both singly and in combination with other supplements. The cleanup level was a little lower than that when only mineral components were added. In the case of the soil with the old residual-oil pollution, the SS and crude oil-stimulants that facilitate the emulsification and availability of a pollutant were needed to enhance this microbial group. Despite the very high concentration of oil products in the soil with the old pollution, the degree of degradation was much higher than it was in the freshly polluted soil, possibly because of the adaptation of the microbial community to the pollution. In the case of the fresh pollution, a correlation was found between HOM number and oil degradation. Such a dependence was less pronounced in the soil with the old pollution. In the oil-polluted samples studied, we found differences in the preferential decomposition of this and other fractions. In the case of the fresh pollution, the maximum decrease in the content of oil products occurred at the expense of degradation of paraffin-naphthene hydrocarbons; in the case of the old pollution, all fractions were equally degraded in the best treatment. The results obtained suggest that each specific case requires its own selection of optimal methods for the enhancement of the microbial communities in polluted soils, in order to accelerate the remediation.

ACKNOWLEDGEMENTS

The work was carried out within the ISTC ## 2093, 1892 and 2488 Projects implementation.

REFERENCES

[1] E.I. Andreuk, G.P. Iutinskaya, P.N. Dulgerov: *Soil Microorganisms and Intensive Land Farming*. Naukova Dumka, Kiev, 1988. 192 p. (in Russian).

[2] V.P. Chekalov: *Method for Determining the Numbers of Hydrocarbon-Oxidizing Bacteria*. USSR No. 1629318. Certificate of authorship, 1991. 3 p. (in Russian).

[3] T.V. Coronelli: Principles and Methods for Intensifying the Biological Degradation of Hydrocarbons in the Environment. *Prikl. Biokhim. Mikrobiol.*, 32 (6), 579 (1996). (in Russian).

[4] J.I. Freijer: Mineralization of Hydrocarbon in Soils under Decreasing Oxygen Availability. J. *Environm. Qual.*, 25, 296 (1996).

[5] J.I. Freijer, H. de Longe, W. Bouten, J.M. Verstraten: Assessing Mineralization Rates of Petroleum Hydrocarbons in Soils in Relation to Environmental Factors and Experimental Scale. *Biodegradation*, 7, 487 (1996).

[6] F.C. Gerretsen: The Influence of Microorganisms on Phosphate Intake by Plant. *Plant and Soil*, 1, 51 (1948).

[7] V.A. Glazovskaya (Ed.): *Restoration of oil-polluted soil ecosystems. A collection of Research Articles*. Nauka Publishers, Moscow, 1988. 254 p. (in Russian).

[8] K.S. Jorgensen, J. Puustinen, A. M. SUORTTI: Bioremediation of Petroleum Hydrocarbon-contaminated Soil by Composting in Biopiles. *Environmental Pollution*, 107, 245 (2000).

[9] C.M. Kao, S.C. Chen, J.K. Liu and Y.S. Wang: Application of Microbial Enumeration Technique Evaluate the Occurrence of Natural Bioremediation. *Water Res.*, 35 (8), 1951 (2001).

[10] N.A. Kireyeva: *Microbiological Processes in Oil-Contaminated Soils*. Bashkirstate University Press, Ufa, 1994. 172 p. (in Russian).

[11] S.G. Malakhov: *Temporary Methodological Guidelines for Monitoring Soil Pollution. Part 2. Moscow Branch of Gidrometeoizdat Publishers,* Moscow, 1984. 55 p. (in Russian).

[12] R. Margesin, F. Schinner: A Feasibility Study for the in situ Remediation of a Former Tank Farm. *World Journal of Microbiology and Biotechnology*, 15, 615 (1999).

[13] R. Margesin, A. Zimmerbauer, F. Schinner: Monitoring of Bioremediation by Soil Biological Activities. *Chemosphere*, 40, 339 (2000).

[14] G.S. Muromtsev: Methods for Studing the Dissolution of Calcium Phosphates by Microorganims. *Mikrobiologiya*, 26 (2), 172 (1957). (in Russian).

[15] S. Rahman, T.J. Rahman, Y. Kourkoutas et al.: Enhanced Bioremediation of n-alkane in Petroleum Sludge Using Bacterial Consortium Amended with Rhamnolipid and Micronutrients. *Bioresour. Technol.*, 90 (2), 159 (2003).

[16] A.V. Ryl'Kov: *Methods for Analyzing the Organic Substance of Rocks, Oil, and Gas. Proceedings*. Issue 122. Western Siberian NIGNI Press, Tyumen,, 1977. (in Russian).

[17] N. Vasudevan, P. Rajaram: Bioremediatuion of Oil Sludge-Contaminated Soil. *Environmental Pollution*, 26, 409 (2001).

[18] X. Wang, R. Bartha: Effects of Bioremediation on Residues Activity and Toxicity in Soil Contaminated by Fuel Spills. *Soil Biol. Biochem.*, 22, 501 (1990).

[19] V.P. Singh, R.D. Stapleton (Eds): *Biotransformations: Bioremediation Technology for Health and Environmental Protection.* Issue 36. Elsevier, Amsterdam, London, New York, Oxford, Paris, Shannon, Tokyo, 2002. 614 p.

[20] B.A. Yagodin, I.P. Deryugin, Yu. P. Zhukov et al. (Eds): *Practical Work in Agrochemistry.* Moscow, Agropromizdat Publishers, 1987. 512 p. (in Russian).

[21] N.S. Yegorov (Ed.): *Manual on Microbiology.* Moscow State University Press, Moscow, 1976. 307 p. (in Russian).

[22] Z.M. Yermolenko, V.A. Chugunov, V.N. Gerasimov et al.: Influence of Certain Environmental Factors on the Survivability of Introduced Bacteria that Degrade Oil Hydrocarbons. *Bioteknologiya*, 5, 33 (1997). (in Russian)

In: New Research on the Environment and Biotechnology ISBN 1-60021-042-2
Editors: A. Kuznetsov and G. Zaikov, pp. 51-56

Chapter 5

ECOLOGICAL-AND-TOXICOLOGICAL ASSESSMENT OF SOIL BIOREMEDIATION PROCESSES

G. A. Zharikov*[1]*, M. S. Sokolov and N. R. Dyadishchev
Research Centre for Toxicology and Hygienic Regulations of Biopreparations (RCT and HRB), Ministry of Health of the Russian Federation, Bld. 102A. Lenin Street, Serpukhov., p/o Dashkovka, Moscow region, 142253 Russia;

"Ensuring chemical and biological safety is a key line of strengthening the national safety of the Russian Federation"
Vladimir V. Putin
(Essential principles of public policy
to ensure chemical and biological safety of Russian
Federation for the period up 2010 and further)

ABSTRACT

A number of countries still have no normative documents regulating the introduction of microorganisms-degraders in the environment or criteria for biotechnology's safety assessment. Our expertise on the implementation of bioremediation technologies of soil contaminated by toxic chemical substances within the framework of the International Science and Technology Center (the ISTC) projects and publications review allows us to develop a number of basic criteria for the process' assessment. Principle criterion of processes and technology's acceptability are for the safety of involved personnel and the health care of the population living in the area of bioremediation. Soil bioremediation technology could be recommended for wide large-scale check and following practical use, for considering health care and ecological-and-toxicological requirements as well as imperative remediation criteria.

Key words: soil bioremediation, microorganism-degraders, ecological-and-toxicological requirements, disease incidence risk for population.

[1] Phone/fax: +007 (0967) 75-97-38; E-mail: toxic@online.stack.net

Pollution of the environment by industrial waste and substances used in agriculture is a serious threat to population health and the nature. Soil self-purification processes could be rather slow because of high pollutant concentration in the environment and, so, active human intervention is required.

It seems to us that the technology for polluted soil biological remediation has some advantages as compared with wildly used physical-and-chemical methods [1]. The advantages are ecological safety of bioremediation processes, no toxic waste left after xenobiotics processing, low laboriousness and low cost.

Bioremediation technologies have to meet rather strict demands. Principle criterion of processes and technologies acceptability is their safety for involved personnel and health care of the population living in the area of bioremediation. At the same time, no internationally acknowledged complex ecological and hygienic norms for a number of soil and soil conjugated environment pollutants is a serious obstacle for practical use of such biotechnologies both in Russia and abroad. Thus, existing in Russia sanitary-and-hygienic and ecological norms regarding soil (minimal permissible level, maximal permissible concentration, approximately permissible concentration, etc.) are only developed for persistent organic pollutants (POP) – polychlorinated biphenyls, dibenzodioxines and dibenzofurans (PCBs, PCDDs, PCDFs), polycyclic aromatic hydrocarbones (3 and 4-benzapilene), and also some pesticides including most persistent ones, chlororganic substances in particular [2].

Selectively developed norms for particular hydrobionts, mesofauna and useful entomoacarifauna species do not guarantee qualitative environment monitoring and maintenance during bioremediation.

Specialists already face a number of difficulties during contaminated site delimitation. It is well-known that mistakes made while soil sampling could cause false results and wrong samples assessment even when most rigorous analysis techniques are employed. Nevertheless, no unified rules for sampling representative specimens of contaminated soil have been laid down yet. Such rules must regulate precisely processes of individual samples taking, their partitioning, packing and transportation as well as prior-analysis short-time storage conditions. Packaging materials and sample labeling procedures must also be standardized. Nowadays, the mentioned requirements accepted in different countries contradict each other in various standards.

Important work on unification and standardization of the mentioned techniques is carried out at ISO (EU international standardization organization) and EPA (Environment Protection Agency, USA). Thus, EPA developed national requirements for various environment specimens sampling, transportation, short-time storage conditions and further study of samples, soil in particular, chemical, biological, microbiological and biochemical methods and biotesting [3]. EPA staff conducts personnel training followed by certificate issuing at 3 specially equipped training centers operating on a regular basis. Such centers are not available in Russia and specialists have to follow all-Union State Standards and departamental regulations of the former USSR in their everyday work [4-5]. Russian sampling requirements are drastically different from the US ones. Thus, the US standards require repeated soil sampling and separate analysis of individual samples, which provides possibility for statistical treatment of obtained results. According to Russian standards an "envelope" soil sampling technique, when 5 individual soil samples are taken and mixed to obtain proportionated sample, is conventional [6].

A number of countries still have no normative documents regulating introduction of microorganisms-degraders in the environment and criteria for biotechnologies safety assessment. Conservancy ministries and organizations (environment protection, agricultural, preserves, water inspection, fish-farms, hunting grounds), apprehending possible side effect, “a priori” make unnecessarily high demands to bioremediation technologies, which slows down their practical use.

In view of the aforesaid we consider development of complex (sanitary-and-hygienic and ecological-and-toxicological) norms to be rather acute task as they are necessary for both microorganisms – pollutant degraders and used biotechnologies safety assessment.

Sanitary-and-hygienic regulations must include:

- Detailed examination of every technical solution,
- Assessment of test microorganisms-degraders harmlessness for warm-blooded animals and human,
- Providing grounds for acceptable type (1-4) of individual protective clothing used by staff,
- Providing grounds for special safety measures for the staff, working in polluted area,
- Minimal acceptable level of soil and soil conjugated environment decontamination,
- Risk assessment for permanent residents of the work area.

Ecological-and-toxicological regulations must include:

- Assessment of test bioagents (microorganisms-degraders, plants, sorbents, etc.) effect on the components of reclamation area environment,
- Assessment of pollutants degradation products integral toxicity,
- Soil phytotoxicity assessment before and after bioremediation,
- Study of soil saprophytic microflora and soil biological activity before and after treatment,
- Prediction of periods (after microorganisms-degraders introduction) necessary for structure and microbial community restoration up to non-contaminated soil value,
- Evaluation of ground ecosystem state by rate indices of transformation and major biogenic elements (carbon, oxygen, nitrogen, etc.) emission.

Our practical expertise implementation of contaminated by toxic chemical substances soil bioremediation technologies within the International Science and Technology Center (the ISTC) ## 228, 2093, 1892 and 2488 projects framework and publications review allow us to analyze the situation and develop a number of basic criteria for the processes assessment. The proposed criteria are approved by some Russian and foreign individual experts and ecological organizations; their implementation provides acceptable efficiency, safety and ecological compatibility of contaminated soil bioremediation.

1. Microorganism-pollutant degraders must be harmless for human. Degrader strains isolated from natural and (or) man-caused areals must not pose any hazard for involved staff. Degraders must be neither pathogenic (i.e. cause no diseases at any human organism penetration route) nor toxinogenic. Harmlessness of selected degrader strains is assessed on conventional laboratory animal models – white mice and rats following most rigorous criteria, internationally used for safety assessment of industrially cultivated microorganisms–drug producer strains. Experimental animals care and use meets international GLP requirements [7]. Strains-degraders harmlessness for warm-blooded animals is studied following [8] and considering WHO guidelines [9] for virulence, toxicity, toxigenicity and dissemination in animal internal organs and tissues.

2. Before conducting the bioremediation procedure, a precise large-scale map of the to-be-remediated site's actual contamination is made. For this purpose, at the first stage, based on the data of chromatographic and chromato-mass-spectrometric analyses of the contaminated soil, spatial and profile distribution of the agent is assessed. When the contaminant composition is unknown, soil analysis is carried out according to the scheme for "analysis of unknown-composition sample" [10] with following composing a list of priority contaminants necessary for bioremediation. In addition, a global soil-agrochemical analysis of the contaminated territory is conducted for determining the type, mechanical composition, sorption properties, nutrition conditions and other characteristics of soil. Orographic features of the territory are represented on map by either traditional methods (horizontals) or using the characteristics of the relief plastics [11]. At the second stage, in the points of potentially maximal accumulation of contaminant, its content and concentration, as well as composition of accompanying and (or) secondary pollutants is determined and the level of the contaminant migration in soil profile and accumulation in water horizon is analyzed.

Analysis of literature on toxic agents-polluted soil bioremediation shows that such theoretical-practical studies should be conducted for quite a long time (2-5 years). Only after the soil contamination is comprehensively studied, it is possible to initiate microbial bioremediation. In parallel with detoxication processes, soil samples are taken to study chemical processes in soil.

In the third year of the studies, residual concentration of the contaminant in soil, as well as the degree of sorption on soil agglomerates and distribution in soil and ground water horizons are assessed.

This allows full and reliable assessment of the observed bioremediation processes and determines the role of microorganisms-degraders.

3. Microbial bioremediation is conducted taking into account the actual micro-zone "pollutant density" on the contaminated territory; this provides for 2- or more repeated precise local introduction of strains-degraders into maximally contaminated micro-zones at the sites under remediation. This helps considerably decrease the cost of bioremediation. Simultaneously, soil samples are taken for evaluation of chemical and biofunctional activity of soil. In addition, residual concentration of the pollutant in soil horizons, the level of sorption by organomineral colloids and localization in the ground water horizon is evaluated, thereby providing a full and reliable assessment of bioremediation processes.

4. Decrease of integral soil toxicity is a critical index of the bioremediation technology efficiency. Existing methods of instrumental chemical analysis do not allow unbiased evaluation of integral (summarized) soil toxicity. This is connected both with the absence of standards (MPC, APC, ASLE) for permissible contents of a number of chemical compounds in soil, and with frequently occurring "cocktail" effect, when chemical compounds interacting with each other are able to strengthen or weaken a joint toxic effect.

For assessment of integral soil toxicity before and after bioremediation, it is necessary to perform bioassay on animal tests (daphnia, earthworms, insects, aquarium fishes, microorganisms, etc). The bioassays assure assessing soil toxicity in the shortest possible time and therefore active monitoring of bioremediation processes.

In addition to the integral toxicity study, phytotoxicity for test plants (oats, radish, watercress, corn, etc.) germs is assessed.

In some instances, for controlling potential danger of transformers or secondary metabolites of pollutants, soil toxicity is evaluated on warm-blooded laboratory animals (white mice and rats).

5. Normalization of biological and biofunctional activity of soil is a critical index of bioremediation efficiency. In this connection, after bioremediation the indices of cleaned-up soil biological activity (total microbial contamination, fermentative activity, etc.) are evaluated in comparison with similar characteristics of control (not-contaminated) soil. Decrease of the microbe quantity in aboriginal soil microflora following introduction of microorganisms-degraders demonstrates their antagonistic interrelations and inhibition of biological processes. On the contrary, activation of soil microflora in the course of the bioremediation accelerates the contaminants degradation. The levels of dehydrogenase and catalase activities are also significant evidence of the processes in soil.

6. An obligatory stage of bioremediation is justification of the degree of the resident population infection risk lowering. Presence of people in bioremediation zone considerably toughens the requirements to medical harmlessness of the applied biotechnology. It is necessary to assess the resident population disease risk. The estimates are made in compliance with techniques US EPA "RISK ASSESSMENT" [12-14]. It should be proved that the use of the biotechnology not only has no toxic effect on people, but also decreases the risk of disease. Otherwise, the biotechnology may not be recommended for the further practical use. For these purposes, the experts of Sanitary Control Services create an electronic database on the health status of the resident population. Then, based on the data from chemical analyses, assessment of the contaminant concentration lowering in soil and, by calculation method, probable decrease of the toxicant concentration in the grown agricultural products is done. Based on the data obtained using the EPA techniques, mathematical calculations are carried out on risk assessment.

As a result, strong evidence must be presented that bioremediation of soil using microorganisms-degraders leads to decrease of the territory contamination and toxicity, and considerably lowers the resident population disease risk.

Therefore, the technology for soil bioremediation is recommended for a large-scale industrial testing and for the following practical use only in case of meeting the approved sanitary-hygienic and ecological-and-toxicological requirements, and imperative remediation criteria, in compliance with which innovative biotechnologies are:

a) Actually harmless for the working personnel and resident population,
b) Efficiently decrease soil contamination and restore the inherent soil functions,
c) Assure obtaining standard-purity agricultural plant and animal production.

References

[1] D.S. Orlov, L.K. Sadovnikova, I.N. Lozanovskaya: *Ecology and preservation of biosphere at chemical pollution;* Moscow, 2002. pp. 300-305.

[2] Materials of UNEP experts sub-regional meeting '*Chemical substances legislation, concerning, in particular, persistent organic pollutants (POP) reduction and destruction*", Moscow, 2001, p. 213.

[3] G.S. Fomin, A.G. Fomin: *Soil. Quality control and ecological safety monitoring in compliance with international standards*. Moscow, 2000, p. 298.

[4] GOST 17.4.3.01-83 Nature preservation. Soil. *General requirements for soil sampling.*

[5] GOST 17.4.3.03-85 Nature preservation. Soil. *General requirements for pollutants assessment methods.*

[6] *Unified requirements for agricultural products, foodstuff and environment objects sampling for pesticides microquantity assessment*: Moscow, 1980, p. 46.

[7] *Good laboratory practice for nonclinical laboratory studies*. 21 cfr part 58 - Food and Drug Administration. 1987.

[8] "Sanitary rules for arrangement, equipment, and maintenance of experimental biological clinics (vivariums)" *the Order of the USSR Ministry of Health #755,* August 12, 1977., the USSR Ministry of Health Guidelines ## 4263–87, ## 2620–82.

[9] WHO Guidelines arrangement, equipment, and maintenance of experimental biological clinics (vivariums): *Bulletin*. WHO, 1981, v. 59, # 6, pp. 20–27.

[10] A.I. Kryliv: Emergency medicine, *journal*, 2001, 3 (10), pp. 32-37.

[11] I.N. Stepanov: *Space and Time in Science about Soil*, Moscow, 2003, p. 184.

[12] *Managing Ecological Risks at EPA. Issues and Recommendations for Progress*: EPA/600/R–94/183 September 1994.

[13] "*RISK ASSESSMENT" PCBs: Cancer Dose–Response Assessment and Application to Environmental Mixtures*: EPA/600/P–96/001F, 1996.

[14] *Proposed Guidelines for Ecological Risk Assessment:* EPA/630/R–95/002 B August 1996.

In: New Research on the Environment and Biotechnology ISBN 1-60021-042-2
Editors: A. Kuznetsov and G. Zaikov, pp. 57-68

Chapter 6

ANAEROBIC AND AEROBIC BIOCONVERSION OF AZO DYES AND THEIR BREAKDOWN PRODUCTS

Natalia Yemashova*[1] *and Sergey Kalyuzhnyi
Department of Enzymology, Chemistry Faculty, The Lomonosov Moscow State University, 119992, Vorobiovy Gory, Moscow, Russia;

ABSTRACT

Acid azo dyes (Acid Orange 6, Acid Orange 7, Methyl Orange and Methyl Red) were completely decolourised in the presence of anaerobic granular sludge, while only Methyl Red was degraded in aerobic conditions using a conventional activated sludge. The experiments with culture broth devoid of cells showed that anaerobic decolourisation of azo dyes was performed by extracellular reducing agents produced by anaerobic bacteria, which was further confirmed by abiotic experiments with sulphide and NADH. The presence of redox mediators like riboflavin led to dramatic acceleration of anaerobic biodecolourisation process. The azo dye reduction products were found to be aromatic amines. Anaerobic toxicity assays showed that the azo dyes were more toxic than their breakdown products (aromatic amines), except 1-amino-2-naphtol. In the presence of activated sludge, only anthranilic acid was completely mineralised while sulphanilic acid was persistent. 4-aminoresorcinol, 1-amino-2-naphtol and N,N-dimethyl-1,4-phenylenediamine underwent autooxidation in aerobic conditions yielding coloured polymeric products. On the contrary, in the presence of granular methanogenic sludge, 4-aminoresorcinol, 1-amino-2-naphtol and anthranilic acid were quantitatively methanised, sulphanilic acid was partially (70%) mineralised while N,N-dimethyl-1,4-phenylenediamine was persistent. Anaerobic toxicity and biodegradability of 2 anthraquinone dyes (Acid Green 25 and Acid Green 27) and triarylmethane dye (Acid Blue 83) were tested. Anthraquinone and triarylmethane dyes were more toxic than azo dyes. Only Acid Blue 83 was biodegraded by anaerobic sludge while Acid Green 25 and Acid Green 27 were persistent.

[1] Phone: (095)9395083; Fax: (095)9395417; E-mail: svk@enz.chem.msu.ru

Keywords: Azo Dye, Anthraquinone Dye, Triarylmethane Dye, Aromatic Amine, Anaerobic Biodegradation, Aerobic Biodegradation

INTRODUCTION

Azo dyes – aromatics substituted with azo groups (-N=N-) – represent a major class of all colorants used worldwide. Antraquinone dyes constitute the second most important class of textile dyes. Presence of such compounds in the industrial wastewater may create serious environmental problems (Hildenbrand *et al.*, 1999). Biological methods of organic pollution removal are commonly considered to be the most economically effective and environmentally sound. In spite of the fact that recalcitrance and toxicity of azo dyes may make biological treatment (especially aerobic one) difficult (Razo-Flores *et al.*, 1997), a number of researches demonstrated azo dye's reduction to colourless aromatic amines by anaerobic microbial consortia (Kalyuzhnyi and Sklyar, 2000; O'Neill *et al.*, 2000; Rajaguru *et al.*, 2000; Tan, 2001; Yemashova *et al.*, 2004; van der Zee *et al.*, 2001). Meantime, a wide range of azo dye structures decolourised in various anaerobic environments allows to assume that decomposition of azo dyes by splitting of their azo bonds is a common property of any anaerobic sludge. At present time, there is no generic opinion concerning the mechanism of azo dye decolourisation under anaerobic conditions. Generally speaking, reduction of azo dye may imply different mechanisms like enzymatic, mediated, pure chemical, intracellular and extracellular (Stolz, 2001; van der Zee *et al.*, 2001, 2003). From the other hand, aromatic amine released from azo dye decomposition are known or suspected to be carcinogens for human (Hildenbrand *et al.*, 1999). Meantime, aromatic amines can be easily biodegraded aerobically through hydroxylation and ring opening of the aromatic moiety. Consequently, it is suggested to combine the anaerobic cleavage of the azo dyes with the aerobic biodegradation for the amines formed (Kalyuzhnyi and Sklyar, 2000; O'Neill *et al.*, 2000; Rajaguru *et al.*, 2000; Tan, 2001). However, such an approach has serious drawbacks. Many of the aminoaromatics formed during the anaerobic decolourisation of azo dyes (especially with hydroxyl groups) are unstable under aerobic conditions and are very prone to autooxidation yielding recalcitrant polymeric products (Stolz, 2001). To avoid this, special attention should be given to a possibility of degradation of such compounds by anaerobic microorganisms. Some researches showed that aromatic amines substituted with hydroxyl or carboxyl groups could be degraded under anaerobic conditions (Battersby and Wilson, 1989; Kalyuzhnyi *et al.*, 2000; Razo-Flores *et al.*, 1997). The objective of this paper was to obtain a deeper insight into mechanisms of anaerobic decolourisation of 4 selected azo dyes (Acid Orange 6, Acid Orange 7, Methyl Orange and Methyl Red) as well as on anaerobic biodegradation of their breakdown products (aromatic amines) and compare the results obtained with the corresponding aerobic biodegradation data. Also anaerobic toxicity and biodegradability of 2 anthraquinone dyes (Acid Green 25 and Acid Green 27) and triarylmethane dye (Acid Blue 83) were estimated.

MATERIALS AND METHODS

Biomass and Basal Medium

Methanogenic granular sludge (16 g VSS/l, aceticlastic activity – 0.19 g CH_4-COD/g/VSS/l) from the EGSB-reactor treating brewery wastewater (Efes-Moscow) and activated sludge (6 g VSS/l) from Kur'yanovskaya municipal aeration station (Moscow) were used for experiments. The basal medium in all batch experiments contained (mg l^{-1}): $NaHCO_3$ (5000), NH_4Cl (280), $CaCl_2{\cdot}2H_2O$ (10), K_2HPO_4 (250), $MgSO_4{\cdot}7H_2O$ (100), yeast extract (100), H_3BO_3 (0.05), $FeCl_3{\cdot}4H_2O$ (2), $ZnCl_2$ (0.05), $MnCl_2{\cdot}4H_2O$ (0.05), $CuCl_2{\cdot}2H_2O$ (0.03), $(NH_4)_2SeO_3{\cdot}5H_2O$ (0.05), $AlCl_3{\cdot}6H_2O$ (2), $NiCl_2{\cdot}6H_2O$ (0.05), $Na_2SeO_3{\cdot}5H_2O$ (0.1), 36% HCl to adjust the pH value to 7.

Assays

The toxicity of the azo, anthraquinone and triarylmethane dyes at concentrations of 150-1200 mg/l and aromatic amines at concentrations of 100-300 mg/l was determined using aceticlastic methanogenic activity assay (Razo-Flores *et al.*, 1997). These assays were performed in 120 ml closed glass serum bottles containing 50 ml of basal medium. After flushing with argon, the aqueous solution of investigated dyes and granular sludge (1.0 g VSS l^{-1}) were added. 0.5 ml of acetate was added from a solution containing 2.85 M sodium acetate. The assay bottles were incubated at 30°C for several days. The concentration of CH_4 and pressure in the headspace of the serum flasks were recorded throughout the incubation. Specific aceticlastic activities of the sludge were calculated from a linear segment of kinetic curves of methane production. The aceticlastic activity of the untreated sludge with a toxicant was taken as a control. From the concentration dependence of the specific activities, the 50% of the inhibiting concentration (IC_{50}) values for the dyes and aromatic amines were estimated. The assays were performed in duplicate.

The anaerobic biodegradation assays with dyes and aromatic amines were performed in 120 ml closed serum bottles filled with 50 ml of basal medium, granular sludge (10 vol. %) and 0.2-1 mM corresponding substrate. Since some biodegradable organics was introduced with the sludge, no any other primary electron donors were added. Moreover, in the preliminary experiments, it was shown that their addition (e.g., ethanol) practically did not enhance biodegradation rates (data not shown). The headspace of the bottles was flushed with argon. The assays were performed in duplicate at 30°C. The concentrations of dyes, aromatic amines, methane, carbon dioxide, sulphide and ammonium were measured. To take into account a production of methane from the cells autolysis, control experiments were carried out without aromatic substrates. Since the concentration of azo dyes may decrease due to adsorption onto the sludge particles, the experiments with autoclaved sludge (killed cells) were performed for checking this effect. In experiments with riboflavin, the latter was added to the standard biodegradation assays at concentrations of 5-15 μM (sludge concentration – 5 vol. %).

The aerobic biodegradation assays with azo dyes and aromatic amines was carried out at 27°C in 100 ml shake flaks filled with 40 ml of basal medium, activated sludge (10 vol. %)

and 0.2-0.9 mM corresponding substrate. The assays were performed for 30 day using thermostatic rotary shaker New Brunswick Scientific (USA) with 180 rpm. The concentrations azo dyes and aromatic amines were monitored.

The chemical azo dyes reduction assays were performed identically as the anaerobic biological assays with the exception that sulphide was added to final concentrations of 5, 0.5 or 0.05 mM instead of granular sludge (sterile conditions). Concentrations of dyes were 0.5, 0.25 or 0.025 mM, respectively. In the assays with reduction by NADH, concentration of azo dyes was 0.25 mM and concentration of NADH was 0.5 mM.

For elucidation of azo dyes decolourisation mechanisms, the assay with culture broth devoid of microorganisms was performed. The culture broth after 10 days of anaerobic biodegradation test was passed through syringe microfilter 0,20 μm (Corning Inc, USA) using for microbial broth sterilisation and then was put into the sterile closed 15 ml flaks. In this way, anaerobiosis was not disturbed. The flasks were flushed with argon, cultivation was performed at 30°C for 40 days and concentrations of azo dyes tested were 0.17-0.4 mM.

Chemicals. All azo dyes and 1-amino-2-naphtol were purchased from Acrus (USA), 4-aminoresorcinol, N,N-dimethyl-1,4-phenylenediamine, anthranilic acid, sulphanilic acid anthraquinone and triarylmethane dyes were purchased from Aldrich (USA). All chemicals were of the analytical grade and were not further purified. All aromatic substrates used are presented in Fig 1.

Analysis. The methane and carbon dioxide contents in the headspace of the serum flasks were determined on a gas chromatograph LHM-8MD (Yagat, Russia) equipped with a steel column (2 m) packed with Porapak QS. The temperature of the column, injector port and conductivity detector was 50°C. Argon was used as a carrier gas with a flow rate of 30 $ml \cdot min^{-1}$. The volume of gas sample was 200 μl. The dyes concentrations were measured spectrophotometrically with a UV-1202 spectrophotometer (Shimadzu, Japan) at maximum absorbance wavelength (pH 7.0) of 428, 431, 465, 490, 555, 590 and 620 nm for Acid Orange 6, Methyl Red, Methyl Orange, Acid Orange 7, Acid Blue 83, Acid Green 27 and Acid Green 25 respectively. Samples were centrifuged and then aliquots of 100 μl were diluted in a 2 ml 0.1 M phosphate buffer (pH 7.0) solution and measured in a 1.0 cm quartz cuvette. The aromatic amines formed were analysed with high performance liquid chromatography (HPLC). The reverse phase column was packed with a Diasorb 130-C_{16}T and had a film thickness of 6 μm. The chromatograph (Gilson, France) was equipped with a UV detector operated at 225 nm. Methanol with 1% acetic acid was used as a liquid phase with a flow rate of 1 $ml \cdot min^{-1}$. A pressure was 14 MPa and the volume of aqueous sample was 10 μl. Ammonium and sulphide were measured spectrophotometrically as described by Fawcett and Scott (1960) and Truper and Schlegel (1964), correspondingly.

Results and Discussion

Biodecolourisation of Azo Dyes.

In the presence of activated sludge, only Methyl Red has become colourless whereas the other azo dyes were persistent during 25 days of aerobic incubation (Fig 2a). This fact may be explained by reduced electron density of the aromatic ring caused by electron-withdrawing azo group making difficult electrophilic oxygenase attack – characteristic for aerobic microorganisms.

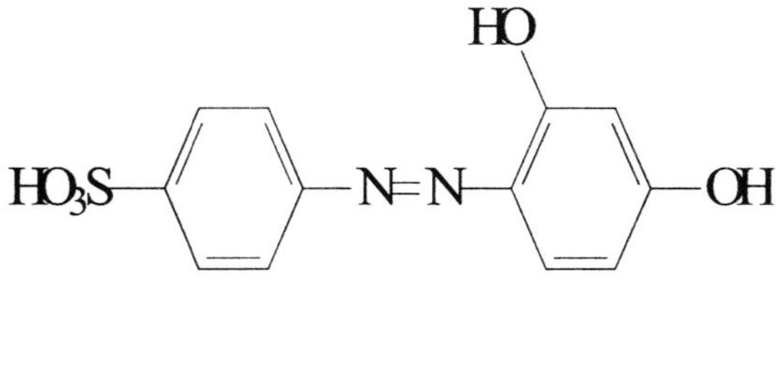

Acid Orange 6

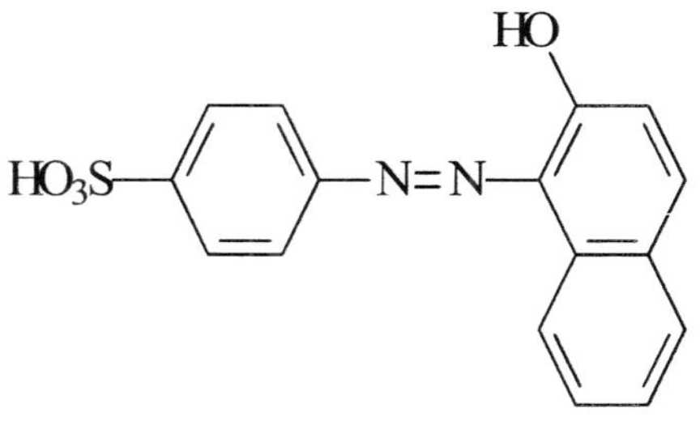

Acid Orange 7

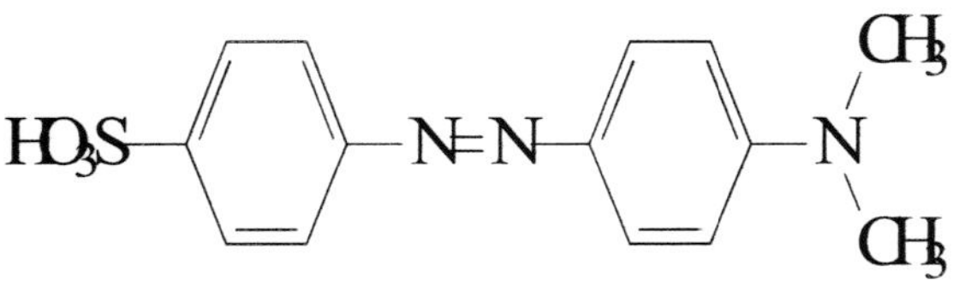

Methyl Orange

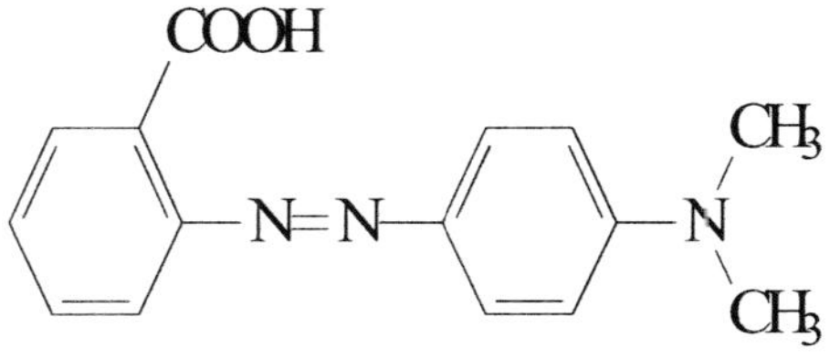

Methyl Red

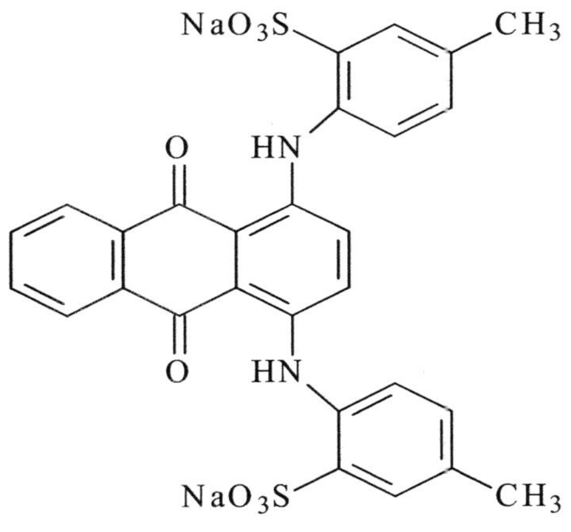

Acid Green 25

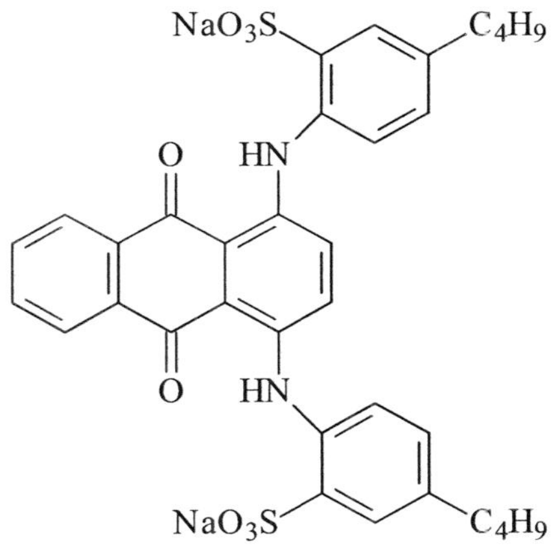

Acid Green 27

Figure 1 continued on next page

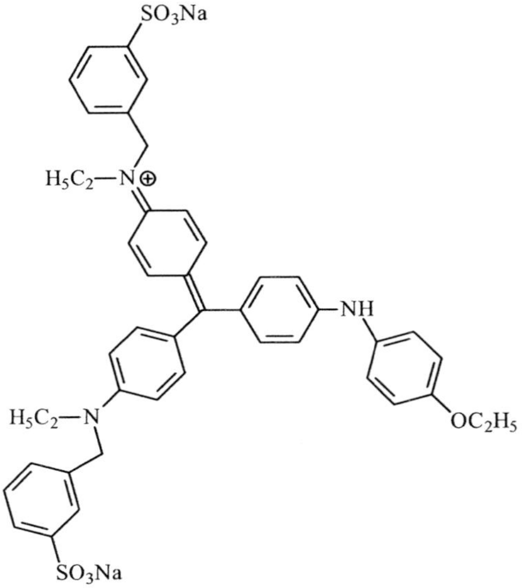

Acid Blue 83

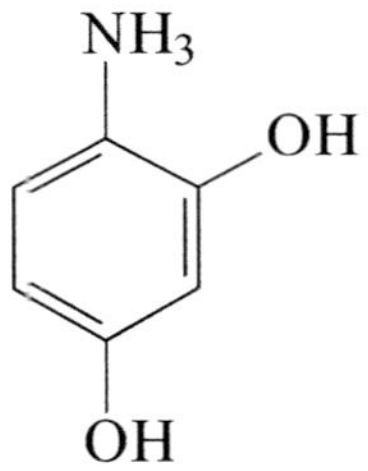

4-aminoresorcinol

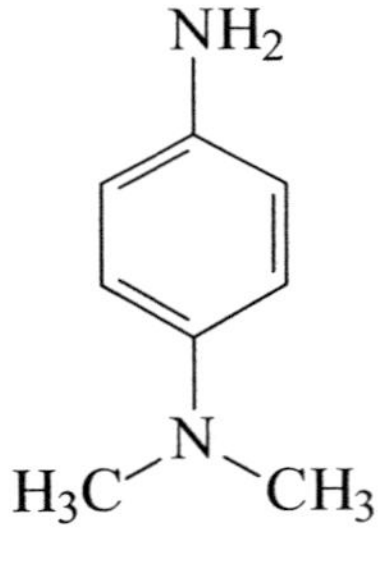

N,N-dimethyl-1,4-phenylenediamine

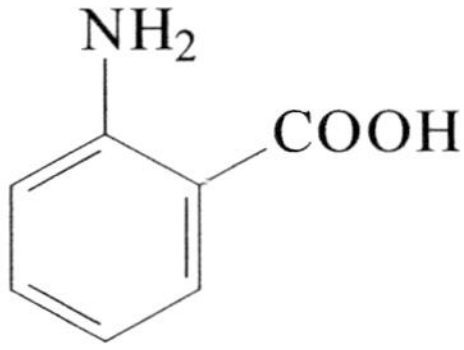

anthranilic acid

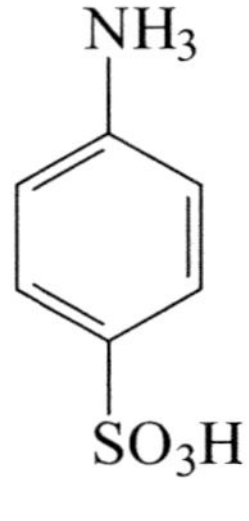

sulphanilic acid

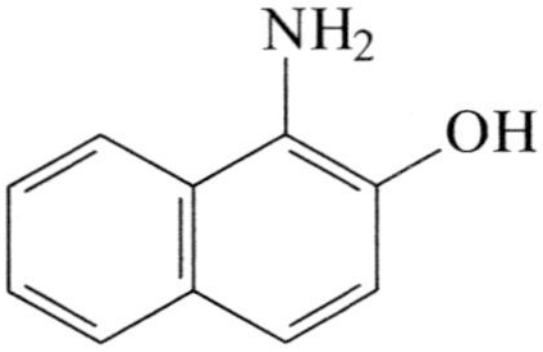

1-amino-2-naphtol

Fig. 1. Structural formulas of N-substituted aromatic substrates used.

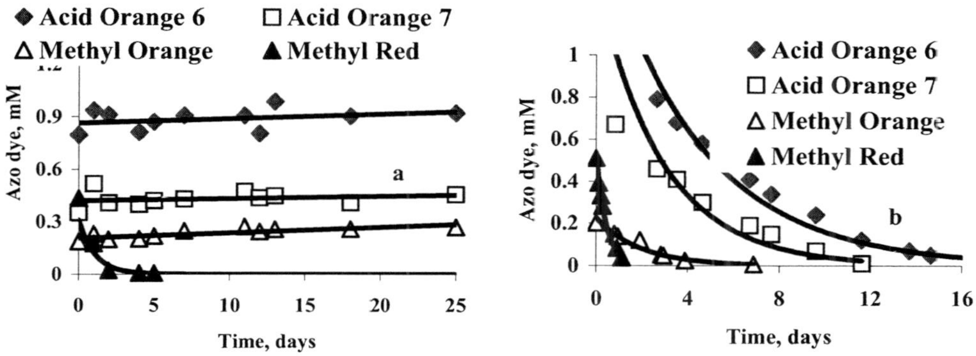

Fig. 2. Aerobic (a) and anaerobic (b) biodegradation of selected azo dyes

On the contrary, in the presence of granular methanogenic sludge, all the azo dyes tested were completely decolourised under anaerobic condition (Fig. 2b). Azo dyes become colourless as a result of reductive splitting of azo group with formation of aromatic amines:

$$Ar^1\text{-}N{=}N\text{-}Ar^2 + 4[H] \rightarrow Ar^1NH_2 + H_2N\text{-}Ar^2 \qquad (1)$$

The bioreduction of azo dyes proceeded without significant lag periods and followed the first-order kinetics (Yemashova *et al.*, 2004) with respect to dye concentration (Fig 2b). The corresponding constants were calculated from the data of Fig. 2b and their values presented in Table 1. It is seen that Methyl Red and Acid Orange 6 were the most prone and persistent dyes for anaerobic reduction, respectively. Addition of ethanol as primary electron donor for reaction (1) practically did not enhance azo dye decolourisation rate (data not shown). It should be also noted that incubation of azo dyes with activated sludge in anaerobic conditions also led to their decolourisation with the first order constants comparable to those observed in the presence of methanogenic sludge (Table 1) witnessing that this ability is not an exclusive property of only anaerobic sludges. Azo dye reduction products identified by HPLC were found to be sulphanilic acid and 4-aminoresorcinol - for Acid Orange 6; sulphanilic acid and 1-amino-2-naphtol - for Acid Orange 7; N,N-dimethyl-1,4-phenylenediamine and sulphanilic acid - for Methyl Orange; and N,N-dimethyl-1,4-phenylenediamine and anthranilic acid - for Methyl Red (Table 2).

Mechanism of Anaerobic Azo Dye Reduction

For degradation of azo dyes the vital functions of microorganisms are necessary because no azo dye reduction was found in the presence of autoclaved anaerobic sludge (data not shown). On the contrary, the culture broth obtained by anaerobic sludge liquor filtration through 0.2 μm microfilter (i.e., devoid of cells but contained the products of theirs vital functions) performed a complete decolourisation of all the azo dyes tested (Fig 3a). Thus, splitting of azo dye appears to be a result of action of reducing agents presented in anaerobic culture broth and followed the pseudo first-order kinetics without any lag period (Fig. 3a).

However, the observed constants (Table 1) were lower compared to those in the presence of anaerobic biomass, which thus is beneficial for maintaining the pools of the reducing agents.

Table 1. The 1st order constants (day^{-1}) of azo dye reduction by sludges, broth, sulphide and NADH.

Azo dye	Methanogenic sludge	Broth*	Activated sludge**	Sulphide, 5 mM	Sulphide, 0.5 mM	Sulphide, 0.05 mM	NADH, 0.5 mM
Acid Orange 6	0.20±0.02	0.09±0.01	0.14±0.01	1.07±0.06	0.10±0.01	0.07±0.01	0.10±0.01
Acid Orange 7	0.30±0.03	0.13±0.02	0.32±0.05	1.44±0.09	0.40±0.08	0.29±0.04	0.62±0.06
Methyl Orange	0.49±0.03	0.36±0.03	0.60±0.06	20±1	0.88±0.04	0.20±0.02	0.13±0.01
Methyl Red	2.0±0.2	0.18±0.01	0.48±0.04	3.9±0.2	0.41±0.05	0.26±0.04	0.11±0.01

*Anaerobic culture broth devoid of cells

** In anaerobic conditions

Table 2. Anaerobic toxicity of azo dyes and corresponding aromatic amines

Azo dye	IC_{50} of azo dye, mM	IC_{50} of corresponding aromatic amines, mM	
Acid Orange 6	2.50	sulphanilic acid non toxic*	4-aminoresorcinol non toxic*
Acid Orange 7	0.90	sulphanilic acid non toxic*	1-amino-2-naphtol 0.54
Methyl Orange	0.25	sulphanilic acid non toxic*	N,N-dimethyl-1,4-phenylenediamine 1.26
Methyl Red	0.99	anthranilic acid non toxic*	N,N-dimethyl-1,4-phenylenediamine 1.26

* non toxic until 2mM concentration

Since sulphide and NADH are usually encountered in anaerobic environments, azo dyes decolourisation was studied without any sludge but in the presence of both compounds. It was found that these reducing agents completely decomposed all the azo dyes studied (data for Acid Orange 6 are shown on Fig. 3b as an example). This abiotic azo dyes reduction also followed pseudo first-order kinetics with respect to the dye concentration though some lag-periods were observed (Fig. 3b). The corresponding constants are presented in Table 1. For sulphide, it was found that the rate of reduction increased with an increase of its concentration (Table 1). However, the sulphide concentrations measured in the anaerobic culture broth were always below 0.05mM. This implies that the observed rates of azo dyes reduction by anaerobic sludge (Table 1) cannot be explained by the action of sulphide only, some other reducing agents like NADH and other redox mediators can also contribute to this process.

The importance of presence of redox mediators in anaerobic media for bioreduction of azo dyes was demonstrated by addition of riboflavin to the standard anaerobic biodegradability assay (Fig. 4). It is seen (data for Acid Orange 6 are shown on Fig. 4 as an example) that even such a low riboflavin concentration as 5 μM led to dramatic acceleration of azo dye reduction.

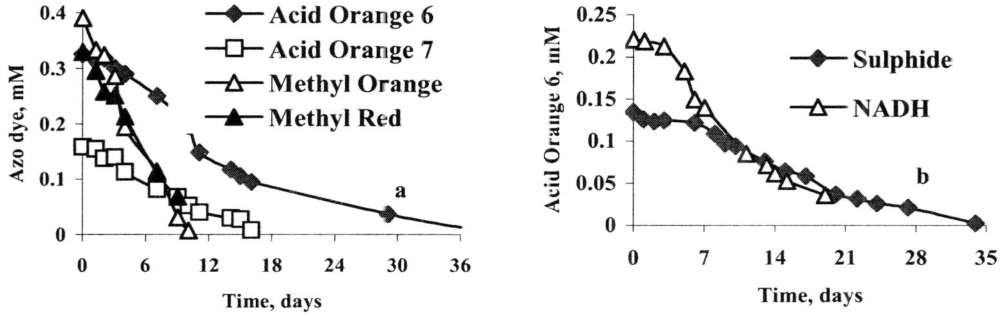

Fig. 3. Decolourisation of azo dyes by anaerobic culture broth devoid of cells (a) as well as sulphide and NADH (initial concentration of both – 0.5 mM) (b).

Anaerobic toxicity of dyes and aromatic amines. Analysis of data presented in table 2 showed that in all cases, except 1-amino-2-naphtol, the azo dyes were more toxic than their breakdown products. This fact is in agreement with literature data when aromatic amines are generally regarded as being less toxic compound for anaerobic microorganisms compared to azo dyes (Razo-Flores *et al.*, 1997). Anthraquinone and triarylmethane dyes were noticeably more toxic than azo dyes (Table 3). Nevertheless dyes concentration in the real waste water is normally lower than anthraquinone dyes IC_{50} value.

Table 3. Anaerobic toxicity of anthraquinone and triarylmethane dyes

Dye	Type of dye	IC_{50}, μM
Acid Green 25	anthraquinone	12.2
Acid Green 27	anthraquinone	7.4
Acid Blue 83	triarylmethane	15.3

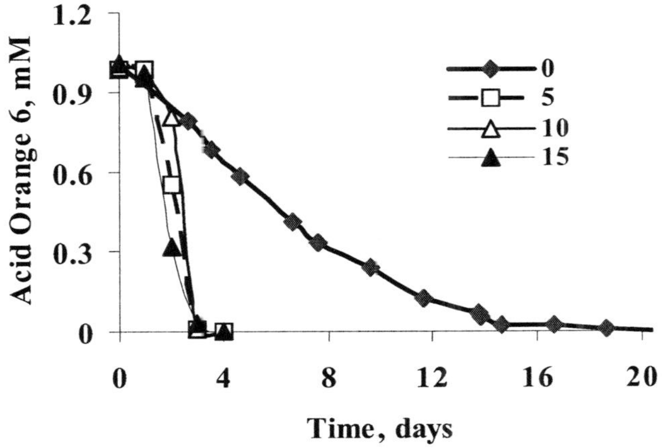

Fig. 4. Influence of riboflavin concentrations (numbers on graph in μM) on Acid Orange 6 anaerobic biodecolourisation (5 vol.% sludge).

Biodegradation of Aromatic Amines

In the presence of activated sludge, only anthranilic acid was completely mineralised while sulphanilic acid was persistent (Fig. 5).

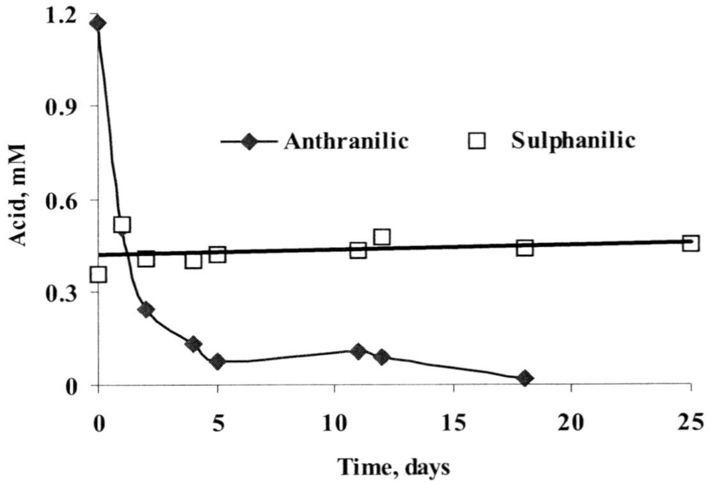

Fig. 5. Aerobic biodegradation of anthranilic and sulphanilic acids

The latter fact is in contradiction to literature data (Tan, 2001) where aerobic biodegradation of sulphanilic acid was observed. Probably, the activated sludge used by us required an acclimation to this substrate. Meantime, 4-aminoresorcinol, 1-amino-2-naphtol and N,N-dimethyl-1,4-phenylenediamine after exposure to aerobic conditions underwent autoxidation and polymerisation yielding coloured products which was evident from a largely altered UV-VIS spectra. Thus, aerobic treatment is not a solution for removal of these amines. On the contrary, in the presence of granular methanogenic sludge, 4-aminoresorcinol (Fig. 6a), 1-amino-2-naphtol (Fig. 6b) and anthranilic acid (data not shown) were quantitatively mineralised to methane, carbon dioxide and ammonium according to the following equations:

$C_6H_7O_2N + 4H_2O \rightarrow 3CH_4 + 3CO_2 + NH_3$ for 4-aminoresorcinol,

$C_{10}H_9ON + 8H_2O \rightarrow 5.5CH_4 + 4.5CO_2 + NH_3$ for 1-amino-2-naphtol

$C_7H_7O_2N + 5H_2O \rightarrow 3.5CH_4 + 3.5CO_2 + NH_3$ for anthranilic acid.

Sulphanilic acid was also slowly degraded in anaerobic conditions producing methane, ammonium and sulphide, however, the initial substrate was not fully mineralised (30% left) after 100 days of anaerobic incubation (data not shown). N,N-dimethyl-1,4-phenylenediamine was not further degraded in anaerobic conditions during 100 days of incubation (data not shown). According to our knowledge, this is the first report demonstrating anaerobic biodegradation of 4-aminoresorcinol, 1-amino-2-naphtol and sulphanilic acid. Two latter findings are in contradiction with some researchers (Tan, 2001;Bras *et al.*, 2001) who reported about non-biodegradability of these substances in anaerobic conditions.

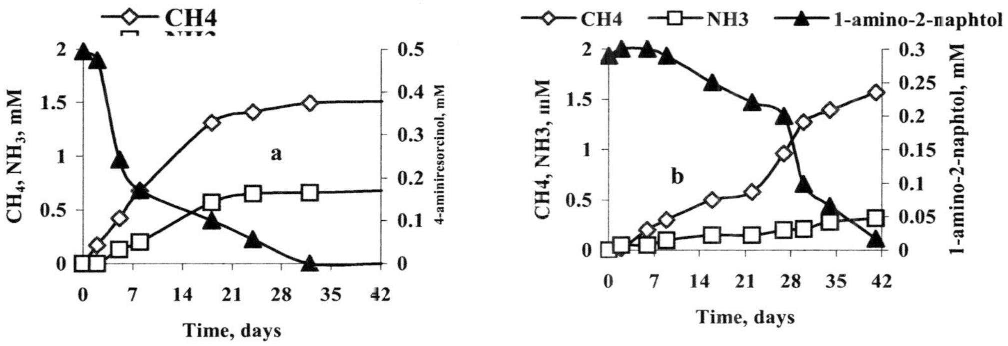

Fig. 6. Anaerobic biodegradation of 4-aminoresorcinol (a) and 1-amino-2-naphtol (b).

Anaerobic Biodegradation of Anthraquinone and Triarylmethane Dyes

In the presence of granular methanogenic sludge only triarylmethane dye Acid Blue 83 has become colourless whereas the anthraquinone dyes were persistent during 100 days of incubation (Fig. 7). During Acid Blue 83 decolourisation no methan or ammonium production was obserwed. Decolourisation of Acid Blue 83 may occure due to reduction of central carbon atom and chromophore structure demage, while no mineralisation was reached. This supposition is in agreement with literature data (Henderson *et al.*, 1997), where transformation of triarylmethane dyes is limited to reversible reaction like reduction of Malachite Green to Leucomalachite Green.

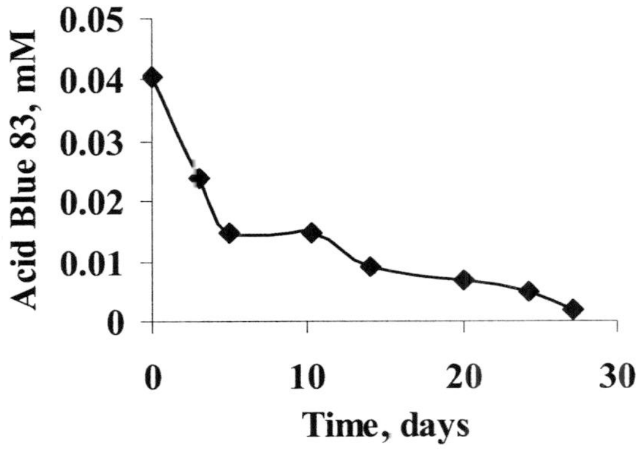

Fig. 7. Anaerobic decolourisation of tryarylmethane dye Acid Blue 83.

Conclusions

[1] All 4 selected azo dyes were completely decolourised by an anaerobic granular sludge, while only Methyl Red was aerobically biodegraded.

[2] The anaerobic decolourisation of azo dyes proceeds due to action of extracellular reducing agents (sulphide, NADH etc) produced by anaerobic bacteria. The presence of redox mediators like riboflavin led to dramatic acceleration of this process.

[3] Anaerobic toxicity assays showed that the azo dyes were more toxic than their breakdown products (aromatic amines), except 1-amino-2-naphtol.

[4] Under aerobic conditions, only anthranilic acid was completely mineralised while sulphanilic acid was persistent. 4-aminoresorcinol, 1-amino-2-naphtol and N,N-dimethyl-1,4-phenylenediamine underwent autoxidation in these conditions yielding coloured products.

[5] Under anaerobic conditions, 4-aminoresorcinol, 1-amino-2-naphtol and anthranilic acid were quantitatively (sulphanilic acid partially) methanised, while N,N-dimethyl-1,4-phenylenediamine was persistent.

[6] Anthraquinone dyes were not biodegraded under anaerobic condition, while triarylmethane dye Acid Blue 83 was transformed by anaerobic sludge to colourless leuco form.

REFERENCES

[1] N. S. Battersby, V. Wilson: *Appl. Environ. Microbiol.,* 55, 433 (1989).

[2] R. Bras, M. I. A. Ferra, H. M. Pinheiro, I. C. Goncalves: *J. Biotec.*, 89, 155 (2001).

[3] J. K. Fawcett, J. E. Scott: *J. Clin. Path.*, 13, 156 (1960).

[4] L. Henderson, T. C. Schmitt, T. M. Heinze, C. E. Cerniglia: *Appl. Environ. Microbiol.*, 63 (10), 4099 (1997).

[5] S. Hildenbrand, F. W. Schmahl, R. Wodarz, R. Kimmel, P. C. Dartsch: *Int. Arch. Occup. Environ. Health,* 72, 52 (1999).

[6] S. Kalyuzhnyi, V. Sklyar: *Wat. Sci. Technol.,* 41 (12), 23 (2000).

[7] S. Kalyuzhnyi, V. Sklyar, T. Mosolova, K. Kucherenko, J. Russkova, N. Degtyaryova: *Wat. Sci. Technol.,* 42 (5-6), 363 (2000).

[8] C. O'neill, A. Lopez, S. Esteves, F. R. Hawkes, D. L. Hawkes, S. J. Wilcox: *Appl Microbiol. Biotechnol.*, 53, 249 (2000).

[9] P. Rajaguru, K. Kalaiselvi, M. Palanivel, V. Subburam: *Appl Microbiol Biotechnol.,* 54, 268 (2000).

[10] E. Razo-Flores, B. Donlon, G. Lettinga, J. A. Field: *FEMs Microb. Rev.*, 20, 525 (1997).

[11] Stolz: *Appl Microbiol Biotechnol.*, 56, 69 (2001).

[12] N. C. G. Tan: Integrated and sequential anaerobic/aerobic biodegradation of azo dyes. *(PhD thesis)* Wageningen University Research Center, the Netherlands, 2001. 104 p.

[13] H. G. Truper, H. G. Schlegel: *Antonie van Leeuwenhoek,* 30, 225 (1964).

[14] N. Yemashova, A. Telegina, I. Kotova, A. Netrusov, S. Kalyuzhnyi: *Appl Biochem Biotechnol.,* 119, 31 (2004).

[15] F. P. Van Der Zee, G. Lettinga, J. A. Field: *Chemosphere*, 44, 1169 (2001).

[16] F. P. Van Der Zee, I. A. E. Bisschops, V.G. Blanchard, R.H.M. Bouwman, G. Lettinga, J. A. Field: *Water Res.,* 37, 3098 (2003).

In: New Research on the Environment and Biotechnology ISBN 1-60021-042-2
Editors: A. Kuznetsov and G. Zaikov, pp. 69-72

Chapter 7

MICROBIAL DETOXIFICATION OF SEWAGE CONTAMINATED WITH OIL, FAT AND PETROLEUM PRODUCTS

A. S. Samsonova[1], Z. M. Aleschenkova and N. F. Syomochkina
Institute of Microbiology, National Academy of Sciences, 220141, Minsk, Kuprevich str., 2, Belarus

ABSTRACT

The ability of microbial cultures *Rhodococcus erythropolis and Rhodococcus ruber* to utilize completely oil/fat and petroleum contaminants of wastewaters in concentrations of 5–10g/l and 1–5.5g/l , respectively, during 96 hours was demonstrated in lab bioreactor.

Key words: microbial degraders, degradation, decontamination, pollution, oil, fats, petroleum products, bioreactor

INTRODUCTION

Disposal of municipal sewage and industrial effluents has always been a top priority problem for human settlements. Nowadays with increased technogenic pressure, especially typical for densely populated urban areas, it has acquired vital significance. Uncontrolled discharge of domestic and industrial wastes complicates the functioning of biological decontamination units detoxifying effluents with activated sludge. Components of activated sludge biocenosis are not able to utilize high concentrations of pollutants. Global practice of microbial treatment of wastewater containing less than 3 % of fatty substances traditionally employs methods of aerobic fermentation using immobilized /1-4/ or non-fixed /5, 6/ bacterial mass of strains-lipid degraders.

---

[1] . E-mail: samsonova@mbio.bas-net.by; fax:+375-172-64-47-66

Efficiency of microbial processes aimed at decontamination of fat/oil-containing effluents indicates economic and technological benefits of applying specialized oil-decomposing strains as compared to physical-chemical or similar procedures.

Accumulation of significant amounts of fatty and petroleum wastes poses grave problems at city aeration station. Domestic sewage, effluents of food industry filled with oils, fats and hydrocarbons are treated by mechanical methods and the resultant piled mass needs to be buried.

Alternative approach disclosed in this paper evaluates efficiency of decomposing fats, oils, petroleum pollutants collected at municipal aeration station by special microbial strains.

Objects and Methods

The objects of studies were: contaminated effluents, and 20 bacterial strains from microbial collection of lab of microbial ecology capable to utilize fatty and hydrocarbon pollutants.

Efficiency of bacterial growth in effluents containing fats, oils, petroleum by-products was evaluated auxanographically on media with agar (20 g agar per 1 l of wastewater). Culture of bacteria able to grow on effluents with oils, fats and hydrocarbons was carried in 500ml Erlenmeyer flasks containing 100 ml of medium, g/l (NaCl - 0,5; $MgSO_4$ $7H_2O$ - 0,8; KH_2PO_4 - 0,7; $(NH_4)_2HPO_4$ - 1,51) supplemented with 1% sewage as C source under aerobic conditions on the shaker (180 rpm) at 29^0C. Each flask was seeded with 10% inoculum.

Biomass production for cultures utilizing, oils, fats and hydrocarbons was performed on meat-wort agar with 0.1 % waste water.

Preliminary assay for lipolytic activity of selected microbial strains was conducted in agar 0.2M phosphate buffer, pH 7.4 (agar –2%) /7/. Lipase activity was estimated titrometrically /8, 9/. Fatty wastes were extracted from acidified waste waters with diethyl ester /10/. Ester extract was washed with water and evaporated in ventilation chamber.

The contents of extracted fatty substances was determined gravimetrically. Petroleum products were identified by IR-spectrometry at Specord M 80 unit with original software.

Protein in waste water was analyzed according to Bradford (26).

Tests on microbial degradation of fats, oils and hydrocarbons were run in 1 liter glass bioreactors at aeration rate 18 l/h. Effluents at municipal aeration station were decontaminated under batch conditions by free-flowing bacterial cultures. Aeration in bioreactor was realized using microcompressor air fed via the valve in the bottom part. Concentration of absolutely dry biomass reached 5 g/l.

Results and Discussion

20 bacterial strains able to utilize oil hydrocarbons and to grow on sewage components sampled at aeration station were chosen from the collection deposited at lab. of microbial ecology. Experiments testing utilization efficiency of effluents containing fats, oils, petroleum have shown active growth of cultures *Rhodococcus erythropolis* and *Rhodococcus ruber*.

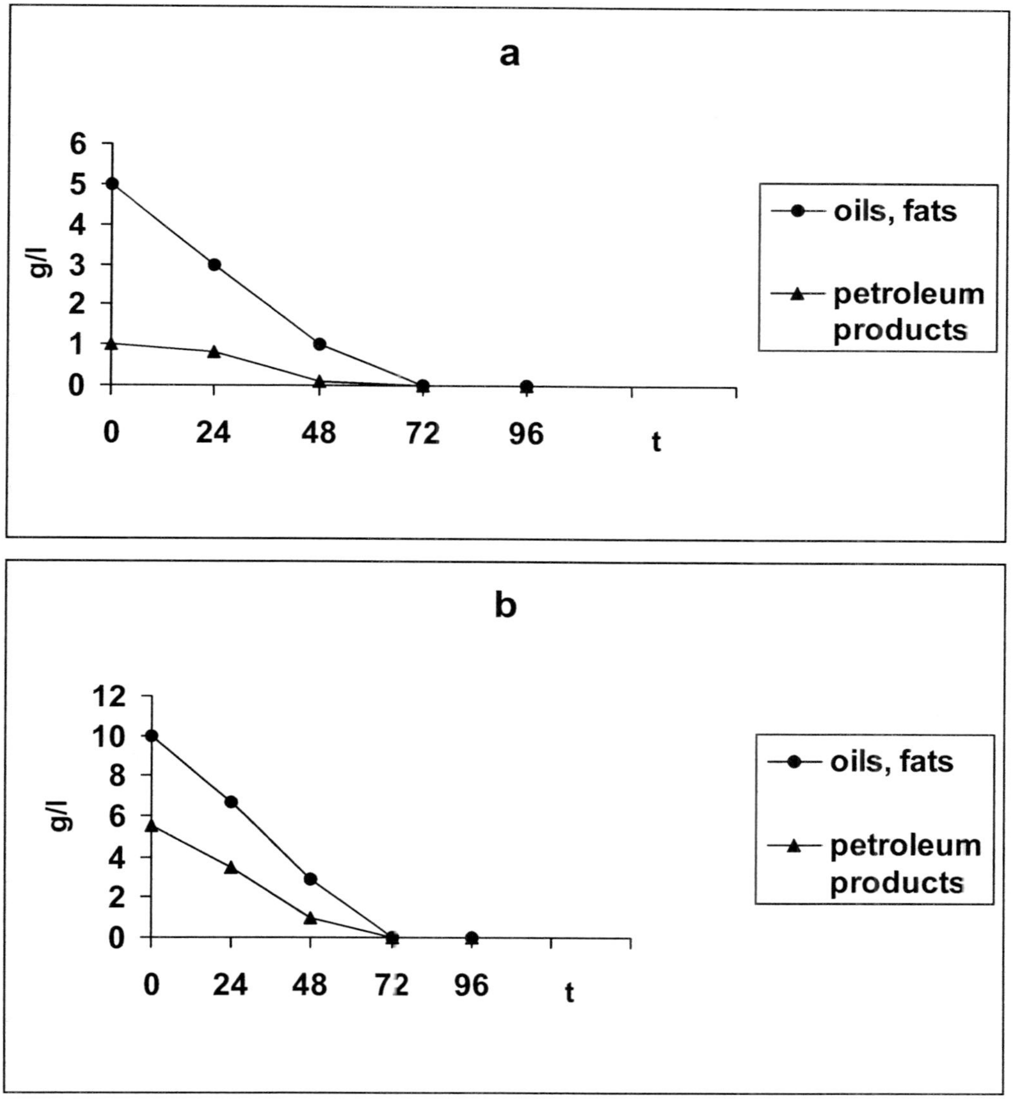

Fig 1. Time course of bacterial degradation of fats, oils and petroleum products in wastewater under lab model conditions

Preliminary estimation of lipolytic activity was performed in selected variants. The results proved presence of lipase in strains displaying good growth on sewage polluted with oils, fats, hydrocarbons. This is evidenced by different in diameter zones of hydrolysis formed around colonies of examined microbial cultures due to activity of bacterial exoenzymes. The highest lipolytic activity was recorded for strains *R. erythropolis* and *R. ruber* forming hydrolysis zones over 15 mm in diameter. For comparison, strains *R. opacus* and *R. maris* distinguished by lower lipase activity yielded more narrow hydrolysis zones (up to 5 mm).

In microbial cultures actively hydrolyzing Tween-80 hydrolysis zones were detected after 24 h of incubation. Cultures less active towards the tested substrate produced hydrolysis zones only by 96 h.

Lipolytic activity of microorganisms utilizing fatty components and oil hydrocarbons as a sole carbon source was revealed in cultural liquid of tested bacterial strains.

Maximal lipase activity among 11 examined microbial strains was registered for *R.erythropolis* and *R. ruber*. Lipase activity of these strains ranged from 0.5 to 0.65 U/mg protein. Less active strains produced lipase at the level 0.1 –0.47 U/mg protein.

Strains *R..erythropolis* and *R.. ruber* were rated as most promising variants for introduction into bioprocesses aimed at removal of fatty and petroleum pollutants from sewage collected at municipal aeration station. Decontamination of wastewaters at municipal aeration station was modeled at lab level in 2 glass bioreactors, of 1 l volume at temperature 15°C and intensive aeration 18 l/h. Wastewaters were treated with dry biomass of *R.erythropolis* and *R. ruber* at the ratio 5 g/l. The decontamination process was carried out in batch conditions during 96 hours. Water containing minimal amount of fats/oils, and petroleum products – 5 g/l and 1 g/l, respectively, was detoxified by 80 % and 90 %, respectively, by 48 h of incubation (fig. 1a).

Fats/oils and hydrocarbons polluting sewage at elevated concentrations – 10 g/l and 5.5 g/l were utilized by microbial degraders in 48 h by 71 % and 81.9 %, respectively (fig. 1b).

After 96 hours complete decomposition of fatty contaminants was observed in bioreactor with minimal load of pollutants. After the same period heavily polluted water was decontaminated to almost 100 % degree.

Thus, ability of microbial cultures *Rhodococcus erythropolis Rhodococcus ruber* to utilize fully oils, fats and petroleum products polluting municipal sewage was demonstrated in lab bioreactor.

REFERENCES

[1] Patent application № 3-16699 JP, 1991.
[2] Patent № 2156749 RU, 2000.
[3] Patent № 19704405 DE, 1998.
[4] Patent № 2787015 JP, 1998.
[5] Patent № 2053204 RU, 1996.
[6] Patent № 2161595 RU, 2001.
[7] J. Kouker, K.E. Jager: *Appl. Env. Microbiol..* **53**, (6), 211 (1987).
[8] I.S. Zviagintseva: *Microbiol.*, **41** (1), 24 (1972) (in Russian).
[9] J. Ota, K. Jamada: *Agric. Biol. Chem.*, **30** (4), 351 (1966).
[10] M.M. Bradford*: Anal. Biochem.*, **72** (1-2), 248 (1976).

In: New Research on the Environment and Biotechnology ISBN 1-60021-042-2
Editors: A. Kuznetsov and G. Zaikov, pp. 73-85

Chapter 8

PREPARATION "RHODER" APPLICATION FOR BIOREMEDIATION OF HYDROCARBON CONTAMINATIONS IN EXTREME CONDITIONS

Valentina P. Murygina[1], Sergey V. Kalyuzhny[1], Patrick Gally[1], Fabian Stirly[1], Liu-Hong[2] and Ouyang-Wei[3]

[1] Department of Chemical Enzymology, Chemical Faculty, Moscow State University, 119899, Moscow, Russia, vpm@enzyme.chem.msu.ru

[2]Department of Environmental Engineering, Beijing University of Aeronautics and Astronautics, Beijing, 100083, China

[3]Beijing Normal University, Beijing 100875, China

ABSTRACT

The bioremediation of oil (hydrocarbon) pollutions, unfortunately, still exists both in the world and in our country, and, accordingly, there is a requirement for the microbial preparations capable to utilize hydrocarbon (HC).

Preparation "Rhoder," developed in the 90's for bioremediation of oil polluted soils and water surfaces (Certificated No 77.99.24.929.D.005978.10.04 by Russian Ministry of Health), was applied with success in Moscow, Western Siberia and Komi Republic in 1994-2002.

In 2000-2004 the preparation "Rhoder" was tested under laboratory and pilot conditions for hydrocarbon decontamination of sand, lawn-and-garden soil, oily sludge which were polluted and heated under 272^{0}C oils, diesel and black mineral oils and wasted heavy oil in sludge from Dongying Oil Production Plant (China, Shangdon province). Moreover the preparation "Rhoder" was used for bioremediation acid (pH=2.4), toxic old heavy oil polluted soil (USA, Montana). As a result, a threefold application of the preparation "Rhoder" showed a decreasein the total hydrocarbon content (THC) of the heated heavy oil at 42-44% (the initial THC was about 5%). After thirteen treatments by the preparation "Rhoder" and phytoremediation (that was between four and five application of the preparation "Rhoder") THC content in lawn-and-garden soil contaminated of crude oil, diesel (each contained ≈10%w) and black mineral (≈1%w) oils decreased at 54-70%, 69-80% and 45-62% respectively. Pilot scale bioremediation experiment with oily sludge showed decreasing of THC content at 56% after the threefold treatment by the preparation “Rhoder” under unfavorable conditions in China, 2004. Whereas composting as an alternative strategy showed decreasing of THC content at 31% only in the same sludge. An insignificant TPH decreasing had observed with activated indigenous microorganisms in the sludge. THC content was decreased at 50-54% after the bioremediation of heavy oil polluted (230g/kg dry weight of soil) hard acidic soil (pH 2.4) from Montana USA by the preparation “Rhoder” in combination with technological know-how. Results obtained show the high oil-degradation potential of preparation “Rhoder”.

Key Words: crude oil, diesel, black mineral oil, oily sludge, preparation “Rhoder”, indigenous oil-degradation microorganisms, bioremediation, composting, phytoremediation

INTRODUCTION

Problem of environment protection from hydrocarbon pollution is one of the main environmental problem in the world. The next problem is a huge amount of oily sludge that are produced during the oil production and processing activities. This sludge usually contains a considerable quantity of heavy oil, which should be removed before its land disposal. The physical-chemical treatments can be applied for the oil pollution and oily sludge, but these methods are extremely expensive [1,2]. Nowadays bioremediation methods are most perspective for this aim. They include both an activation of indigenous microorganisms and introduction of oil-oxidizing ones [3]. Today bioremediaton and composting are two major economic methods for the decontamination of oil pollutions [3,4,5]. Composting has some visible advantages including relatively low capital and maintenance costs, simply design and operation and some (but incomplete) removal of oil pollution. However, in general, the efficiency of composting is unsatisfactory to meet the current environmental regulations. High bioremediation efficiency for oil polluted soils have been demonstrated by numerous researches [6,7,8,9], but these methods have limitations for the oily sludge mainly with extremely high pollution level and recalcitrant contaminants for biodegradation.

Preparation “Rhoder” developed in 90th for bioremediation of oil polluted soils and water surfaces was applied with successes in Moscow region, Western Siberia and Komi Republic in 1994-2002 for restoration of fresh and old oil contamination of soils and water surfaces [7,8,10]. In Western Siberia and in Komi Republic, especially on sites situated behind of Polar Circle, climatic conditions are extreme for bioremediation of oil pollutants, however, this article deals with extreme properties and features of soil and HC that contaminated these objects. Our more than 10 years experience on bioremediation of oil polluted water surfaces and soils shows that the main factor determining success of works, is not only a preparation of oil-oxidizing microorganisms, but also the technological procedures providing optimal conditions for degradation HC by microorganisms used in a preparation.

Also this article deals with the bioremediation of quartz sand and soil polluted by different kinds of oil and petrol’s pollutants; oil polluted soil with extremely characteristics and oily sludge using the preparation “Rhoder” and composting of oily sludge with manure.

Material and Methods

The preparation "Rhoder" (Certificate №.77.99.24.929.D.005978.10.04 issued by Russia Ministry of Health) consists of two oil degrading strains *Rhodococcus* and represents a concentrated suspension of living bacterial cells with 10^9 to 10^{10} CFU/ml, ready for the application without a preliminary activation or dry powder cells with 10^{10} CFU/g [11]. The working suspension of the preparation "Rhoder" for bioremediation contained 10^6-10^7 oil degrading cells per ml + Fertilizer (0.2% solution) N_{60}:P_{60}:K_{90} for laboratory experiments and Carbamide with KH_2PO_4 for the pilot test. Each experiment had a negative control without any treatment, positive control with activation of indigenous microorganisms by addition of fertilizer and own experimental vessels where the preparation "Rhoder" was added periodically.

Laboratory Experiments

Quartz sand, used as a soil model, was washed by water, then dried (105 ^{0}C, 1 h), sieved (0.63 mm) and placed into plastic vessels (15x12x5 cm^3, 600 g of sand/vessel) in threeplicate before the contamination with heated oil (about 4.5%w/w). The oil contaminated vessels were kept under ambient temperature and light during 6 weeks. The contaminated sand was treated three times by the suspension of "Rhoder" and fertilizer. The vessels were periodically watered and mixed throughout the experiment. The soil samples were taken before each preparation and fertilizer treatment.

Lawn-and-garden soil from Moscow region was artificially contaminated with crude oil, diesel (both about 10% w/w) or black mineral oil (about 1% w/w). The soil dried and sieved as described above and placed into plastic vessels of the same size before the contamination. Due to high clay content in this soil, peat (1/3 w/w) or sand (1/2 w/w) was used as a soil-structuring agent. Since the peat had pH 4.6, fast and slow lime was added in the proportion 1:1 (26 g /vessel) and mixed carefully (the added sand did not change pH). All experimental variants were duplicated. After contamination and thorough mixing, the vessels were placed into hothouse at 20-23^{0}C without natural light. The contaminated soils were treated 12 times by the suspension of "Rhoder" and fertilizers. Throughout experiments, the vessels were periodically watered and mixed for intensification of pollutants degradation. Between fourth and fifth treatments with "Rhoder", the phytoremediation was performed during one and a half month with a combination of annual and perennial Gramineae seeds (16.5 g/vessel) previously treated with an anti-stress agent. Phytoremediation was effected under natural light and ambient temperature with regular watering but without sampling. All grasses and ungerminated seeds were pulled out of vessels after the phytoremediation. The final samples were taken to determine the residual hydrocarbons.

A heavily oil-polluted (about 300 g TPH/kg soil dry weight (d.w.)) acidic soil (pH 2.4) from the site Kevin, Montana, USA was remediated by bio-augmentation of the preparation "Rhoder"or by phytoremediation with oat species (*Avena sativa*) in plastic vessels (ϕ14cm x h 5cm). Peat compost was added to the polluted soil at weight ratio 1:1 (for the preparation "Rhoder"; and oat seeds) and 1:2 (for oat seeds). pH was increased to 6.6-6.8 with addition of a lime. No peat compost and lime were added to the negative control soil. Oat seeds were sowed in all vessels except those where the preparation was tested. Bioaugmentation was effectuated by three times application of the preparation "Rhoder". Experiments were led under ambient temperature, periodically watering, mixing soils with preparations. The second phytoremediation with oat seeds *(Avena sativa)* was performed two weeks after finishing the bioremediation and the first phytoremediation. 1/10 w/w of peat-compost were added to each vessel before the second phytoremediation.

The Pilot Experiment

The pilot experiment was performed with oily sludge at Shengli Oil Production Plant, (Dongying, China) under roof on the surface of a concrete slab. There were 4 compartments (2.5x2.0x0.2 m^3) for testing of preparation "Rhoder", 4 compartments (1.6x1.0x0.2 m^3) for positive and negative controls and 1 (0.75x0.5x0.2 m^3) - for composting. The initial concentrations of hydrocarbon (HC) in oily sludge varied from 327.7 g/kg to 371.0 g/kg, HC in old oil contaminated soil was 151.0 g/kg. The oily sludge and oil-polluted soil were very clay and boggy, so straw (8kg) and sawdust (15kg), top sand (100 kg) and clean soils (20 kg) were used as additives to the sludge and soil (table 1). 60 kg crude manure and 10 kg straw were added to the sludge and mixed thoroughly before composting. This pilot experiment was performed under unfavorable conditions. In April 2004 it was rather cold +10-11^{0}C and the temperature inside oily sludge varied from 11^{0}C to +18-20^{0}C and never rised higher during experiment.

Preparation "Rhoder" was applied in concentration of 0.3g dry preparation per 1 kg of modified sludge every two weeks. The modified sludge was mixed after spreading the solution of preparation with fertilizers or only fertilizers for activation of indigenous microorganisms and mixing and watering every 3 days. At the same days when sludge and soil were treated by the preparation "Rhoder", the next 10 kg of crude manure was added to the composting mass. Experiments lasted during 58 days under the ambient temperature. After the end of the treatments, the Tall Fescue lawns (*Festuca arundinace*) were transplanted on the surface of sludge and soil to perform a subsequent phytoremediation of these oily sludge and soil and check their toxicity.

In the laboratory experiments the samples for analyses were taken before each treatment and 1/2-1 month after the last treatment according to 5 points scheme and each sample was examined [12]. In pilot experiment the samples for analyses were taken randomly from the upper layer (5.0 cm of depth) every 3 days and 2 weeks after the bioremediation finishing. THC content was measured by gravimetric, colony chromatography, GC and GC-MS methods. GC was equipped with a flame ionization detector and a 30 m stainless steel column. The operation conditions were as the following: initial temperature was 50°C and increased at 10°C/min to 200°C and then increased at 5°C/min to the final temperature of 290°C. Helium as a carrier gas was delivered at a rate of 0.8 ml/min.

Numbers of microorganisms in the treated soils were estimated by MPN method, using the meat-peptone agar for heterotrophic bacteria (HT) and Raymond's medium with oil for hydrocarbon-oxidizing bacteria (HCO) [13]. Soluble nitrogen and phosphorus in soil, humidity and pH were determined with standard agrochemical methods [14].

RESULT AND DISCUSSION

In the laboratory experiment the quartz sand was contaminated with heating oil without light boiling HC fractions under 272°C. As a result triple treatments by preparation "Rhoder" of the sand polluted by heavy oil decreased the concentration of HC at 42-44 % (the initial HC concentration was 4.3 %). In the positive control (only fertilizers treatments) HC concentration decreased at 12-13 % (fig. 1). HC concentration in the negative control not changed. The analysis of HC groups in heated oil before and after bioremediation showed, that the ratio of alkenes, aromatic and polyaromatic fractions changed. The content of polar compounds that included also metabolites of HC oxidizing microorganisms (fig.2.) essentially increased, that confirmed the biodegradation HC of heavy heated oil by oil-degradation microorganisms of the preparation "Rhoder".

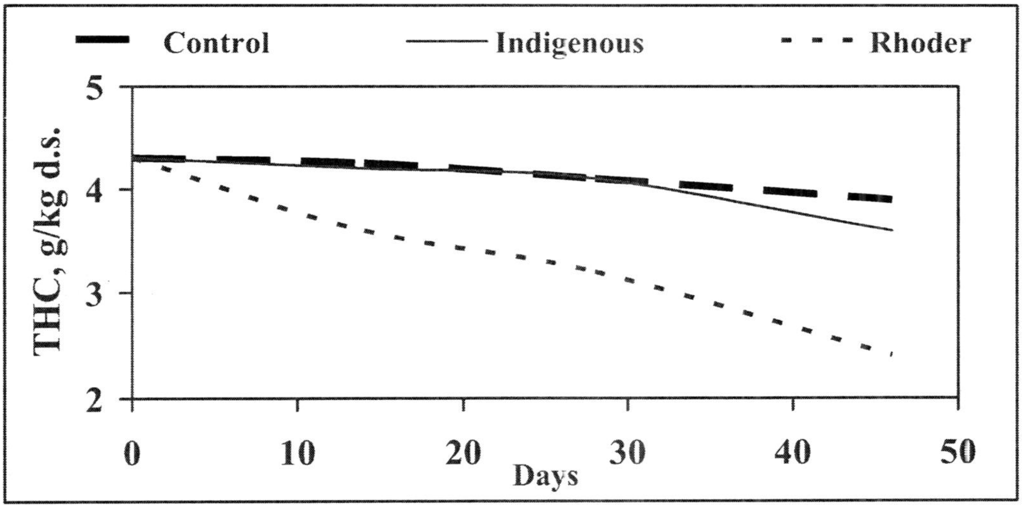

Fig.1 Bioremediation of heated oil polluted sand by the preparation "Rhoder"

Table 1. The composition of the materials subjected for bioremediation during pilot experiment

Variants	Oily sludge, kg	Initial THC, g/kg	Manure kg	Straw kg	Sawdust kg	Top sand, kg	Terminal THC,g/kg
CM	**280**	**126.8**	**60**	**10**	**0**	**0**	**101.4**
Sludge 1	**750**	**328.0**	**0**	**12**	**15**	**0**	**251.3**
Sludge 2	**600**	**371.0**	**0**	**8**	**15**	**100**	**121.2**
Soil *	**800**	**150.0**	**0**	**8**	**15**	**20***	**94.1**

Note: * 20 kg of pure soil were added

The laboratory experiment with lawn-and-garden soil showed that the initial level of crude oil, diesel and black mineral oil decreased at 57-64%, 54-67%, and 72-30%, respectively, after 5 consecutive treatments with "Rhoder" (Table 2.). Phytoremediation decreased the concentration of crude oil and diesel at 10-30% and 15-20 %, respectively, depending on peat or sand used as the structuring agent. No influence of phytoremediation was observed for black mineral oil variants (Table 2).

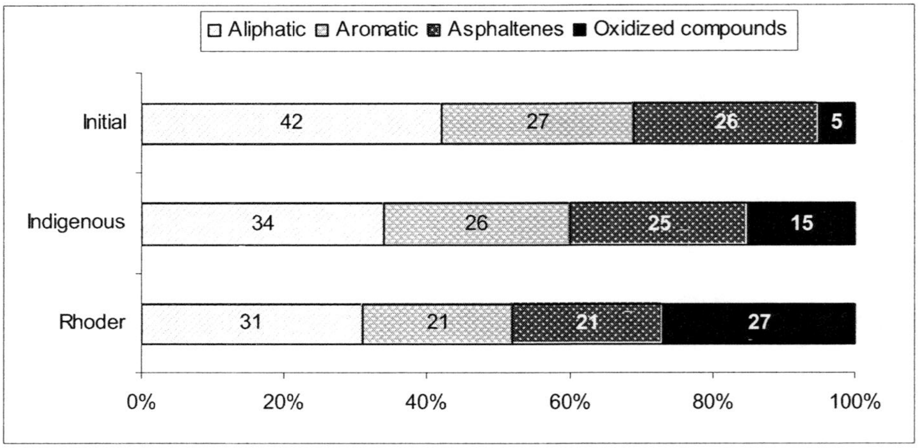

Fig. 2. The analysis of HC groups in heated oil polluted sand before and after bioremediation by the preparation "Rhoder"

Table 2. Degradation of hydrocarbon contamination in lawn-and-garden soil by preparation "Rhoder"

Variants		THC, g/kg, the preparation treatment							Degradation, %
		Before	14 days after 1	14 days after 2	14 days after 3	14 days after 4	After phytoremediation	14 days after 5	
Crude oil, 10%	Peat	33.10	28.81	19.73	19.50	19.78	16.15	14.20	57.1
	Sand	48.43	33.45	37.60	33.50	32.02	15.22	17.52	63.8
	Without treatment	61.26	55.98	55.70	54.90	55.0	54.78	53.80	12.2
Diesel, 10%	Peat	36.03	38.43	31.14	28.11	22.00	16.60	16.58	54.0
	Sand	49.72	49.63	43.37	37.33	30.96	20.64	16.38	67.1
	Without treatment	61.02	65.06	64.70	60.40	57.49	52.10	51.79	15.1
Black Mineral oil, 1%	Peat	11.0	10.09	6.58	3.77	2.78	2.64	3.02	72.5
	Sand	6.39	6.18	4.83	3.32	4.40	5.22	4.27	30.9
	Without treatment	7.50	7.65	7.04	7.39	7.12	7.45	7.00	6.7

Following seven treatments by the preparation "Rhoder" insignificantly decreased the concentration of residual HC in the experiment. Total HC content in lawn-and-garden soil contaminated of crude oil, diesel (each contained ≈10%w) and black mineral (≈1%w) oils decreased at 54-70%, 69-80% and 45-62% respectively (fig.3-5). The preparation "Rhoder" degraded hydrocarbons of crude oil, diesel and black mineral oil efficiently during the first three or four introductions. Then, probably, after the available hydrocarbons have been exhausted, the oil degradation process became slower in spite of substantial quantity of oil-oxidizing microorganisms, soluble nitrogen and phosphorus, optimal humidity and pH in the soil.

During the phytoremediation the growth of plants was better with a peat as the structuring agent. The peat and sand used as the structuring agents for clay HC contaminated soil were equally efficient in the case of bioremediation with the "Rhoder". The peat as the structuring agent showed a high efficacy especially when it was added to the black mineral oil contaminated soil, perhaps, due to humic substances presented in the peat. The peat humic substances are considered to promote assimilation of difficult available polycyclic hydrocarbons [15]. At the same time, sand as the structuring agent of crude oil polluted clay soil increased the level of bioremediation especially with the combination of fourfold "Rhoder" treatments and the phytoremediation (Table 2). Results obtained showed the possibility of the preparation "Rhoder" to degrade diesel and black mineral oil besides of crude oil in lawn-garden clay soil.

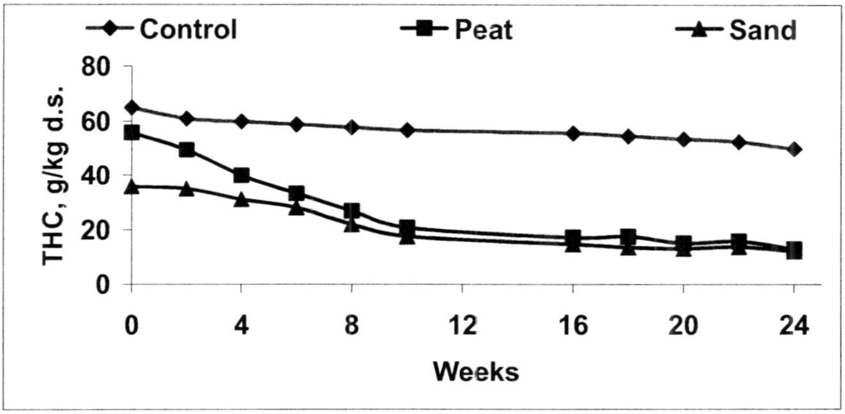

Fig. 3. Bioremediation with preparation "Rhoder" and Phytoremediation (10-16 weeks) of crude oil polluted lawn-and-garden soil

Table 3 Bioremediation of oil polluted soil Kevin by the preparation "Rhoder"

Soil Kevin, rate soil:peat-compost	HC g/kg d.s.		Degradation	
	Initial	Final	%	HC mg/kg d.s. per day
Negative control	306.7	305.5	0.4	41
Rhoder 1:1	228.3	104.6	54.2	4123
Rhoder 1:1	228.3	112.6	50.7	3858

In 2004 the third laboratory experiment was performed in order to find a technology permitting decrease the HC contamination in toxic oil polluted acidic soil. In 2001-2002 "Landfarming" bioremediation technology was performed with a preparation "Rhodart-2000" without success on the site Kevin, Montana, USA. Activation of indigenous microorganisms didn't work as well. It was a heavily oil-polluted (about 300 g TPH/kg soil dry weight) acidic soil (pH 2.4) with extreme characteristics (heavy sour oil with 11.5 % of sulfur and high content of benzothiophene). The laboratory experiment was performed with the soil that was taken from a negative control in 2001-2002. Preliminary laboratory investigations had shown that the initial Kevin oil polluted soil without any treatment was toxic for cells of both preparations "Rhodart-2000" and "Rhoder". It was the cause to modify of "Landfarming" technology used in 2001-2002 in Montana, USA. Peat compost and a lime were added to the polluted soil as described above.

In laboratory experiment we compared the bioaugmentation of the preparation “Rhoder” and phytoremediation using oat species (*Avena sativa*). Obtained results showed that peat compost was a key element to permit phytoremediation and microbial HC degradation of heavily oil polluted Kevin soil (table 3). Bioaugmentation was effectuated by three times application of the preparation “Rhoder” with addition of peat compost in ratio 1:1. THC content decreased at 50-54% with average degradation rates of 4.1 g HC/kg soil d.w. per day. The phytoremediation by using *Avena sativa* with addition of 1/10 w/w peat compost to the soils after preparation “Rhoder” last treatment decreased THC content additionally at 9-15% in soil of Kevin site (table 4).

The first phytoremediation by using *Avena sativa* with addition of peat compost in ratio 2:1 and 1:1 decreased TPH content in soils of Kevin 3 at 37-41% (table 5). The second phytoremediation with *Avena sativa* with addition of 1/10 w/w peat compost to the soil decreased TPH content at 7- 13% additionally (table 6).

The analysis of HC groups in oil polluted soil Kevin before any treatment and after phytoremediation and bioremediation with the preparation “Rhoder” showed, that the ratio of alkenes, aromatic and polyaromatic fractions changed. The content of polar compounds that included also metabolites of HC oxidizing microorganisms (fig.6.) increased more than 2 times. Aromatic fractions decreased more than 3 times that confirmed the biodegradation of HC of heavy oil by the preparation “Rhoder”. After phytoremediation we also observed the ratio of alkenes, aromatic and polyaromatic fractions changing but not so essential.

So, obtained results showed that the addition of peat compost was very important for successful bio- and phytoremediation of oil polluted toxic soil. The microbial HC degradation by the preparation “Rhoder” with a combination to the subsequent phytoremediation yields the best results, than each of these remediation technologies of HC polluted toxic soil.

In a pilot test in China, 2004, oily sludge contained heavy hydrocarbons and all components of stratum water and components which were applied during oil recovery. The part of sludge was dehydrated, but mobile forms of nitrogen and phosphorus were much enough and pH was close to neutral in the sludge. Pilot survey has revealed presence HC oxidizing microorganisms in amount of 1000 CFU/g of dry sludge. The addition of selected oil-degradation bacteria with the preparation “Rhoder” increased the number of HC oxidizing bacteria more than 10^6 times in oily sludge and oil contaminated soil. Under these circumstances the liquidation of oil contamination was more efficient than in positive control compartments. So, the pilot experiment of hydrocarbon decontamination of oily sludge and soil with the threefold preparation “Rhoder” treatments in China, 2004 under unfavorable conditions showed decreasing of THC content at 45-47% and 53%, respectively (Fig.7). At the same time, activation of indigenous microflora with fertilizers (positive controls) decreased the THC in sludge and soil at 13-20 and 23%, respectively. There was no essential THC decreasing in negative controls. After composting, the THC decreased at 31% (Fig. 7)

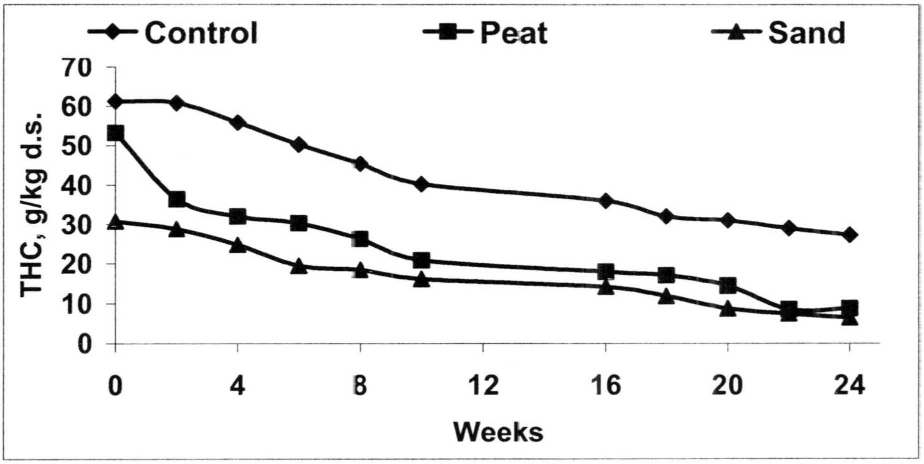

Fig. 4. Bioremediation with preparation "Rhoder" and Phytoremediation (10-16 weeks) of diesel polluted lawn-and-garden soil

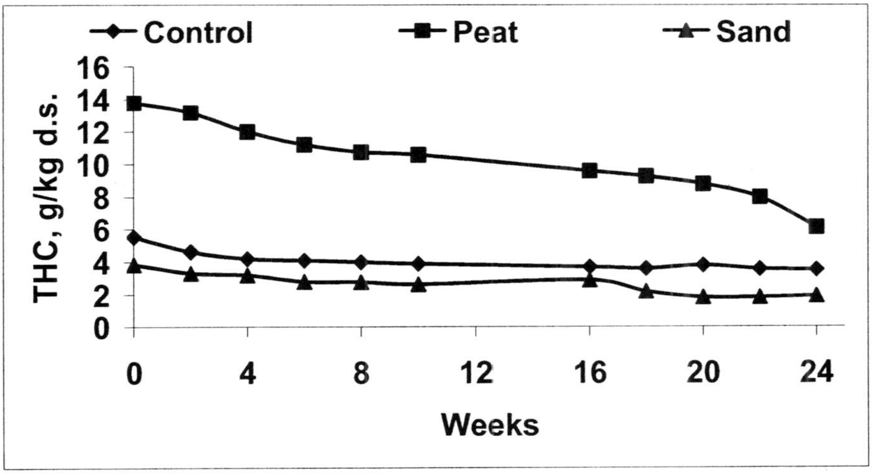

Fig.5 Bioremediation with preparation "Rhoder" and Phytoremediation (10-16 weeks) of black mineral oil polluted lawn-and-garden soil

The obtained results showed that the degradation of HC by preparation "Rhoder" was the most rapid during the first month and then the rate declined with time. The planting of Tall Fescue (*Festuca arundinace*) decreased THC content at 5-7% additionally and showed a low toxicity of sludge after bioaugmentation of the preparation "Rhoder" and after composting.

The gas chromatograph analysis showed (fig. 8) that alkenes in the range of C13- C29 were the major compounds in the initial oil pollution. They were mainly linear ones and C18 content was at most - 5.3%. The smaller peaks between the linear alkenes (Fig. 8) indicates a presence of the branched-chain alkenes and the most part of them contained less than four methyl groups according to the MS data. The main branched compound was phytane (2,4,10,14-tetramethylhexadecane, $C_{20}H_{42}$) - 10.9%. The GC data for treated samples also showed an enhanced oil degradation by the preparation "Rhoder" compared to the CM - the peaks of n-alkenes (n<18) in the CM were taller than in sludge No3. The branched alkenes are more resistant to microbial degradation than n-alkenes due to their molecular structure. Furthermore, larger molecular structure are considered to be less degradable than smaller ones [16,17]. Our data confirm these observations.

So, the efficiency of preparation "Rhoder" application in comparison with composting for decontamination of oily sludge has an evident proof.

Table 4. Phytoremediation of oil polluted soil Kevin by *Avena sativa* after bioremediation with the preparation "Rhoder"

Soil Kevin	THC,g/kg d.s.		Degradation	
Soil:peat-compost ratio	Initial	Final	%	THC mg/kg d.s. per day
Negative control	305.5	291.0	4.7	690
Rhoder, 10:1	104.6	95.4	8.8	438
Rhoder, 10:1	112.6	95.4	15.3	819

Table 5. The first Phytoremediation of oil polluted soil Kevin by *Avena sativa*

Soil Kevin	THC,g/kg d.s.		Degradation	
Soil:peat-compost ratio	Initial	Final	%	THC mg/kg d.s. per day
Sand	4,6	-	-	-
Fresh soil, Kevin	7,0	-	-	-
Negative control	306,7	305,5	0,4	41
2:1	255.2	152.4	40.3	3429
2:1	255.2	160.3	37.2	3164
1:1	216.2	128.6	40.5	2921
1:1	216.2	126.4	41.5	2992

Table 6. The second phytoremediation of oil polluted soil Kevin by *Avena sativa*

Soil Kevin	THC,g/kg d.s.		Degradation	
Soil:peat-compost ratio	After the first phytoremediation	Final	%	THC mg/kg d.s. per day
Sand	4,6	-	-	-
Fresh soil, Kevin	7,0	-	-	-
Negative control	305,5	291.0	4.7	690
2:1	152.4	140.5	7.2	523
2:1	160.3	147,7	7.8	573
1:1	128.6	111.9	12.8	785
1:1	128.6	111.9	12.8	785

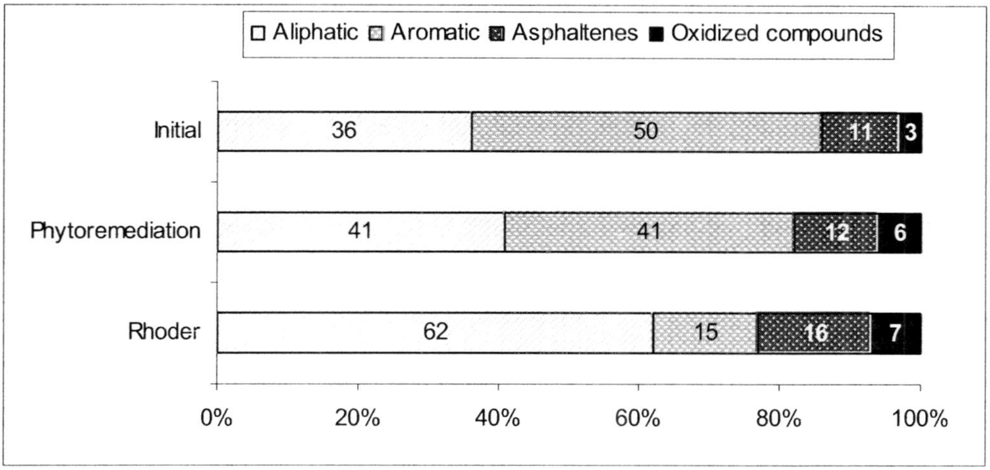

Fig. 6. The analysis of HC groups in oil polluted soil Kevin after phytoremediation with *Avena sativa* and bioremediation by the preparation "Rhoder"

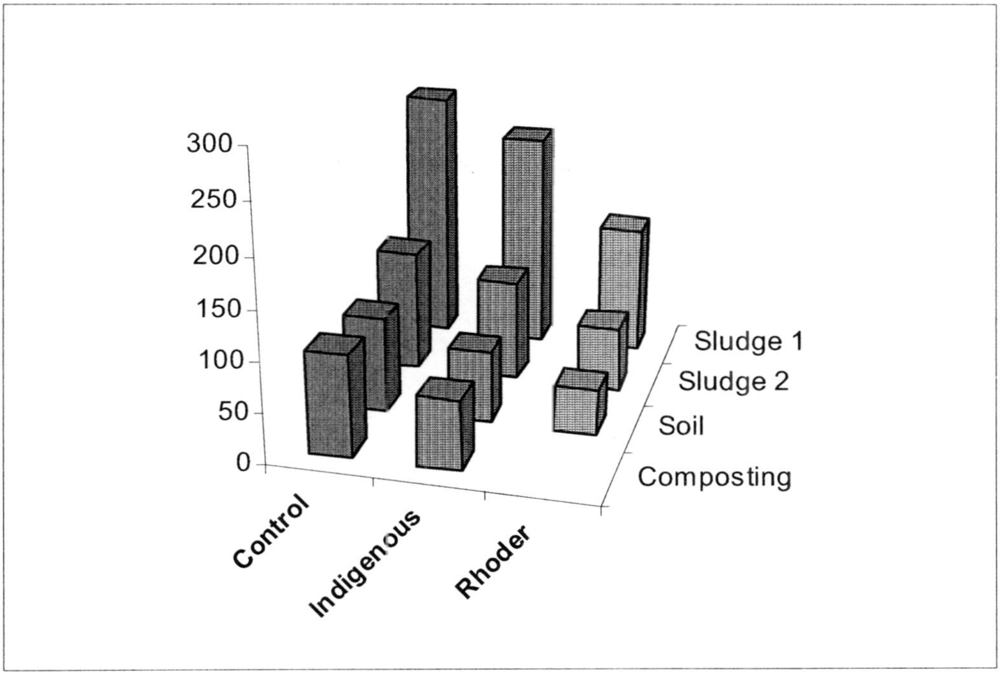

Fig. 7. Bioremediation of oily sludge and old oil polluted soil by the preparation "Rhoder" in comparison with composting

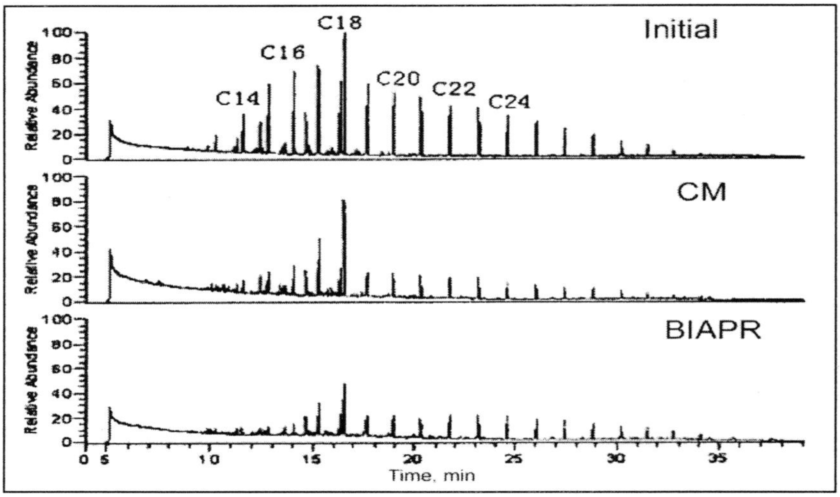

Fig. 8. The gas chromatograph analysis of oily sludge before any treatment, after Composting (CM) and after bioremediation by the preparation "Rhoder"

Conclusions

The obtained results show a high hydrocarbon-degradation potential for the preparation "Rhoder" under different pollutants and extreme soil characteristics, and also the importance of using different technological procedures to provide an effective work of the preparation. The results also showed positive influence of phytoremediation which should be carried out after bioremediation with the introduction of the preparation that contained oil-degrading microorganisms.

References

[1] Ed. D. Sheehan: *Bioremediation Protocols*, Humana Press, 339 (1997)

[2] E. Rosenberg, E.Z. Ron, In *Bioremediation: Principles And Applications* (R.L. Crawford, D.L. Crawford, Eds.), Cambridge University Press, New York, USA, 100 (1996)

[3] I. Lazar, S. Dobrota, A. Voicu: *Journal Of Petroleum Science And Engineering*, Volume: 22, Issue: 1-3:15, 160 (1999).

[4] V.P. Murygina, S.V. Kalyuzhnyi, M.U.Arinbasarov: *Biodegradation*, 11 (6), 385 (2000).

[5] Namkoong, Wan; Hwang, Eui-Young; Park, Joon-Seok: *Environmental Pollution*, 8: 23~31 (2002)

[6] B.K. Gogoi, N.N. Dutta, P.Goswami: *Advances In Environmental Research*, 6:767-782 (2003)

[7] V.P. Mouryguina, S.V. Kalyuzhnyi: *Proc. Of The First European Bioremediation Conference,* 3-5 June, Crete, Greece, 28 (2001).

[8] V. Mouryguina, S. Kalyuzhnyi: *In Proc. Of The Second European Bioremediation Conference*, 30 June-4 July, Crete, Greece, 127 (2003)

[9] M.A. Simon, J.S. Bonner, A.Ch. Page, et al. : *Ecological Engineering*, 7: 263-277, (2004).

[10] V.P. Murygina, M.Yu. Markarova, S.V. Kalyuzhnyi: *Environment International,* V.31, No 2,163 (2005).

[11] V.P. Murygina, N.E. Voishillo, S.V. Kalyuzhnyi: Patent RF No 2174496 (2001).

[12] A. Alexandrov, V. Zagvozdkin, G. Ertzev, et.al.: *Proc. Inetrn. Confer. Of ICS And UNIDO "New Technologies For Clean Up Oil Contaminated Waters And Soils, Remaking And Use Oil-Slimes"*., 10-11 Of December, Moscow, 96 (2001) (In Russion).

[13] T. Nazina, Ye. Rozanova, S. Belyayev, M. Ivanov: *Preprint* Biological Center Press, Pushchino, 1988. 35 P. (In Russian)

[14] E. Durynina, V. Egorov: Msu, Moscow, 147 P (1998) (In Russian).

[15] K.M. Salem, I.V. Perminova, N.Ju. Grechischeva, V.P. Mouryguina, S.V. Mescheryakov: *Ecology And Industry*, April, 19, (2003) (In Russian).

[16] H. De Jonge, J.I. Freijer, J.M. Verstraten, J. Westerveld, F.W.M. Van Der Wielen: *Environmental Science And Technology* 31, 771 (1997).

[17] R.M. Atlas, R. Bartha: *Residue Review* 49, A83 (1973).

In: New Research on the Environment and Biotechnology ISBN 1-60021-042-2
Editors: A. Kuznetsov and G. Zaikov, pp. 87-92

Chapter 9

KINETIC OF BIODEGRADATION OF N-ALKANES BY *RHODOCOCCUS RUBER* AND *RHODOCOCCUS ERYTHROPOLIS*

D. V. Zhukov*[1]*, V. P. Murygina and S. V. Kalyuzhnyi
Enzymology Subdepartment, Chemical department, Lomonosov M.V. Moscow State University, Leninskie Gory 1/11, Moscow, Russia, 119992.

ABSTRACT

This paper describes a study on the kinetic biodegradation of a mixture of n-alkanes (C13-C17) by *Rhodococcus ruber* (Ac - 1513-D) and *Rhodococcus erythropolis* (Ac - 1514-D) that are composed of the oil-degradation preparation "Rhoder". Kinetic characteristics of liquid paraffin's biodegradation by chosen strains (specific growth rate (μ) and specific activity of a biomass (q)) are obtained. It has been shown that two strains in a consortium consume paraffins more actively than each strain individually. Besides, it has been shown that the limit factor of microbial population development is a decrease of pH of cultural broth.

Key words: kinetic of biodegradation, synergism, biodegradation of hydrocarbons, *Rhodococcus spp.*, Rhoder.

The problem with the purification of various natural objects from pollution by crude oil and products of its processing is still actual. In this connection, the biotechnological methods based on the ability of some microorganisms to use hydrocarbons and other components of oil as a sole source of carbon and energy became more widely demand.

Usually communities of bacteria are used for bioremediation. *Rhodococcus spp.* show one of the highest petro-degrading activities. These microorganisms or the other species (*Acinetobacter*, *Actinomyces*, *Arthrobacter*, *Thiobacterium*, *Pseudomonas*, *Bacillus*, *Actinobacter*, and also some kinds of microscopic fungi, for example *Candida sp.* [2,3,4]) with the combination of *Rhodococcus spp.* are components of the biopreparations that are successfully used to remediate the oil-polluted soil and water. In particular, the preparation

[1] E-mail: dimok_zh@mail.ru

"Rhoder" includes two highly active oil-degrading bacteria: *Rhodococcus erythropolis* Ac-1514 and *Rhodococcus rubber* Ac-1513 [1]. However, in-depth studies of kinetic regularities for revealing the influence of every strain separately and the combined influence on components of hydrocarbon substrata have not been carried out.

Now great attention is being paid to study the interaction between hydrocarbon-degrading cultures that are used to obtain the consortium to enhance efficacy, and deep utilization of various substrate complexes. As a rule the efficacy of consortium hydrocarbon-degrading microorganisms is more higher than that of monoculture preparation [5,6,7].

Such a consortium can be certain (received by means of a combination of some strains with known degrading ability), and uncertain (actually - enrichment cultures) with unknown characteristics of each member of the consortium. Each approach has merits and demerits. The first is strictly certain, but to increase a degree of biodegradation and to expand a number of utilized components, the consortium from the big number of strains is necessary. With the second approach - obtaining so-called uncertain consortia can be especially effective for degradation of components of some mineral oils. However in some cases various co-metabolic processes in consortium occur including, for example, accumulation of toxic substances which inhibit the further development of microorganism population [8, 9].

Thus, the purposes of this work were:

- studying kinetic regularities of growth *Rhodococcus ruber* and *Rhodococcus erythropolis* and their metabolism on the mineral medium with a mix of liquid paraffins as a sole source of carbon;
- revealing synergism with both strains during biodegradation of hydrocarbons.
- as a substrate we chose a mix liquid n-alkanes (C13-C17).

MATERIALS AND METHODS

In the present work *Rhodococcus ruber* (Ac-1513-D) and *Rhodococcus erythropolis* (Ac-1514-D) are investigated. Strains have been allocated in laboratory of microbiology VNII Neft since 1991-1992.

Cultivation of Microorganisms

Mineral medium contained (g/l): $CaCl_2*6H_2O$ - 0,01; $MnSO_4*7H_2O$ - 0,02; Na_2HPO_4 - 1,5; KH_2 PO_4 - 1,0; $MgSO_4$ - 0,2; NH_4Cl - 2,0; NaCl - 5,0. pH 7. It was sterilized under 120 °C (2 atm.) during 30 min. Mixture of liquid paraffins was used as a substrate (vol. %): $C_{13}H_{28}$–7,5; $C_{14}H_{30}$–24,0; $C_{15}H_{32}$–30,6; $C_{16}H_{34}$–24,2; $C_{17}H_{36}$–11,2; admixture – 2,5.

Determination of CO_2

The carbon dioxide contents in the gasspace of the flasks were determined on a gas chromatograph LHM-8MD (Yagat, Russia) equipped with a steel column (2 m) packed with Porapak QS. The temperature of the column, injector port and conductivity detector was

50°C. Argon was used as a carrier gas with a flow rate of 30 ml·min^{-1}. The volume of gas sample was 200 μl.

DETERMINATION OF LIQUID PARAFFINES

Extraction of liquid paraffines was carried out by hexane (10 ml + 5 ml + 5 ml). After evaporation of hexane undecane (40 mkl) as internal standard was added to sample. The hydrocarbons contents were determined by capillary gas chromatography equipped with column (Ultra 1, 50m *0,2mm*0,5mkm) packed with crosslinked methylsilicone. The temperature of the column was 120 °C (5 min) to 230 °C (20 min) at 10 °C/min. The temperature of the injector and detector (FID) was 300°C. Helium was used as a carrier gas with a flow rate of 20 ml·min^{-1}.

DETERMINATION OF PROTEINS

The growth of microorganisms was identified by the determination of intracellular protein concentration [10]. An aliquot (1 ml) of culture broth was taken in every 4 hours in exponential phase and less than in 4 hours in the rest of the time. The cells were broken in 1 ml NaOH (3 M) by boiling in the water bath for 45 min. Then assays were centrifuged (12000 r.p.m., 10 min.) again to separate a residue. 3 ml of Bradford reagent (100 mg Cumassi brilliant blue G-250 in 50 ml of 95% ethanol and 100 ml concentrated phosphoric acid, than distilled water was added to get 1 l of Bradford reagent) were added to 100 μl of the preparing assays. The A_{595} was measured after 10 min. passed.

RESULTS AND DISCUSSION

Adaptation of Cultures

The cultures *R. ruber* and *R. erythropolis* was shown to have the lowered degrading activity after storage on beef-extract agar and the subsequent transfer to medium with hydrocarbons of oil as a unique source of carbon. The lag phase for *R. ruber* was about 7 days and for *R. erythropolis* was about 5 days. Therefore the enrichment of these cultures on the selective liquid medium containing liquid paraffins was made. Thus, as the result the enriched *R. erythropolis* Ac-1514D and *R. ruber* Ac-1513D were obtained by means of four serial sub inoculations in specific liquid medium with liquid n-alkanes. The lag phases were reduced to 50 hours for the both cultures and more reduction of lag phases was not shown.

Distinction in growth of enriched strains was observed. *R. erythropolis* generated white suspension in the depth of the medium with hydrocarbon substrate. Opposite, cells of *R. ruber* were located on the surface of hydrocarbon drops because of more lipophilic cell surface of this strain.

STUDYING THE KINETIC REGULARITIES OF BIODEGRADATION OF A MIX N-ALKANES BY ENRICHED R. ERYTHROPOLIS AND R. RUBER

Experiments on studying biodegradations kinetics of liquid paraffins by adapted strains have been led. For that, subculture (0,5 ml) was carried into hermetically closed flasks (inclusive 40ml mineral medium and 200mkl substrate). The initial most probably number (MPN) of cells were averaged 10^6 cells/ml. The results of studying of hydrocarbons degradation kinetics by *R. ruber* and *R. erythropolis* (separately and in a consortium) are represented on figure 1. During experiments the biodegradation of the substrate was averaged at 34% for *R. ruber* and 79% for *R. erythropolis* whereupon processes were stopped spontaneously. It is necessary to note, that equimolar transformations of the substrate to CO_2 in our experiments was not observed. This fact can be explained that the substrate ceases to be accessible for extraction as a result of its adsorption on a cellular wall, its transportation and accumulation in cells.

During the experiment it was shown that the limiting factor of microbe population development was the pH decrease of culture broth. Thus when achievement the value pH 4,75 for *R. erythropolis* and 5,65 for *R. ruber* processes of growth and substrate catabolism were inhibited.It is clear from experimental data (figure 1c) that during the lag-period of the growth of *R. ruber* there was a consumption of the substrate. This fact can be possible explained by the initial adsorption of substrate by cells followed by their biodecomposition. For *R. erythropolis* (figure1a) such effect was not observed. Thus, it is possible to make the conclusion that biological mechanisms of hydrocarbon degradation by strains considered are various. It is connected probably with essential distinction in a chemical composition of each strain cellular wall. *R. ruber* has a more hydrophobic cellular wall. Since influence of two strains on the substrate is different it is probable that a composition of strains will utilize a hydrocarbon substrate more successfully. For the proof of this assumption experiment on study biodecomposition of liquid paraffins by a mixture of two strains was carried out.

It is evident from the obtained kinetic parameters (specific growth rate and specific activity of the biomass) that *R. ruber* has a greater specific growth rate than *R. erythropolis*, but development of *R. ruber* population stops earlier and a specific activity of the biomass considerably lower than that one for *R. erythropolis* (table 1).

For the consortium these processes passes more quickly than for individually cultivated *R. erythropolis* and they are more exhaustive than for *R. ruber*. The quantitative data on specific growth rate of the biomass and its specific activity for the consortium of microorganisms are represented in table 1. It is obvious from tha data on specific activity that two strains in the consortium consumed paraffins more efficiently than every strain particularly. It allows us to conclude that a synergism at the community of *R. erythropolis* and R. ruber is observed. Thus as a result of the addition less active *R. ruber* to more active *R. erythropolis* the specific activity of hydrocarbon biodegradation by the bacterial community was increased.

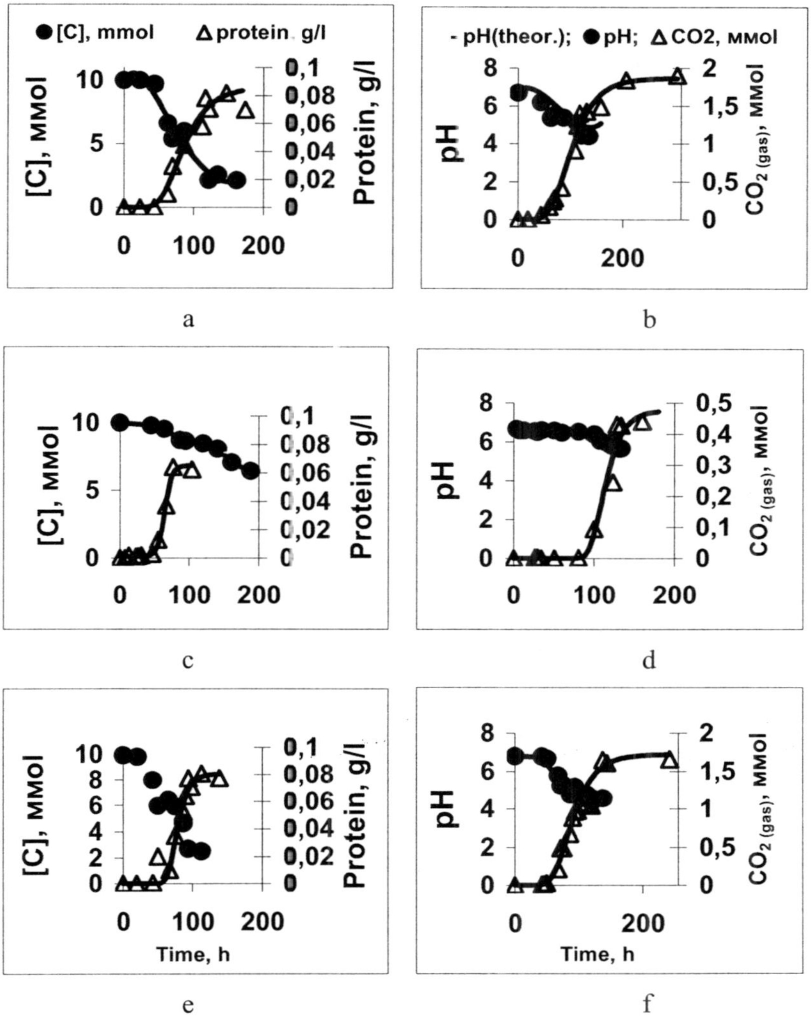

Figure 1. Dynamics of change of sistem parameters (cell proteins, substrate, pH, CO_2) during cultivating of *R. erythropolis*(1a, 1b), *R. ruber*(1c, 1d), *R. erythropolis+R. ruber*(1e, 1f).

Table 1. Kinetic characteristics of the biodegradation of liquid paraffins by *Rhodococcus* spicies

R. erythropolis	R. ruber	R. erythropolis + R. ruber
Specific growth rate, 1/h		
0,032	0,074	0,041
Specific activity of the biomass, mmol[C]/g/h		
77,0	33,6	98,5

CONCLUSION

It was shown in our experiments that the decrease of pH of culture broth was the limit factor of microbe population development. Besides, kinetic characteristics of microbe growth processes and consumption of the substrate by individual strains and their consortium was found. It is clear that addition of less active *R. ruber* to more active *R. erythropolis* leads to increased activity of the hydrocarbon substrate consumption, which is proof of the synergism in biodecomposition of hydrocarbons. Thus, the use of these strains in the community is preferable in carrying out the bioremediation.

REFERENCES

[1] Murygina V.P., Voishvillo N.E., Kalyuzhnyi S.V. *Patent RF No 2174496.*

[2] L. G. Whyte, L. Bourbonniere, C. W. Greer: *Appl. Environ. Microbiol.*, 63 (9), 3719 (1997).

[3] S. Meyer, R. Moser, A. Neef, U. P. Stahl: *J. Microbiology*, 145, 1731 (1999).

[4] D. Koma, Y. Sakashita, K. Kubota, Y. Fujii, F. Hasumi, S.-Y. Chung, M. Kubo: *Biosci. Biotechnol. Biochem.*, 67 (7), 1590 (2003).

[5] T.V. Koronelli: *Usp. Mikrobiol.* (in Russian), 15, 99 (1980).

[6] M. V. Gusev, T. V. Koronelli, Yu. N. Korolev: *Mikrobiologiya* (in Russian), 47 (6), 10 (1978).

[7] F. Jacquot, P. Doumenq, M. Guiliano, D. Munoz, J. R. Guichard, J. Mille: *Talanta*, 43, 319 (1996).

[8] T. V. Koronelli, T. I. Komarova, A. V. Ignatchenko: *Mikrobiologiya* (in Russian), 52 (1), 94 (1983).

[9] T. V. Koronelli, T. I. Komarova, A. V. Ignatchenko: *Mikrobiologiya* (in Russian), 53 (2), 213 (1984).

[10] R. M. C. Dawson, D. C. Eleiott, W. H. Elliot, K. M. Jones: *Data for Biochemical Research*. Mir, Moscow, 1991.

In: New Research on the Environment and Biotechnology ISBN 1-60021-042-2
Editors: A. Kuznetsov and G. Zaikov, pp. 93-97

Chapter 10

AN ANAMMOX-LIKE PROCESS IN THE WASTE WATER TREATMENT OF THE COKE-CHEMICAL PLANT

Fu Yigang[1], A. Ye. Kouznetsov[2,*] and V. A. Stepanenkov[3]
[1]East China University of Science and Technology, Shanghai, China
[2]D. Mendeleyev University of Chemical Technology of Russia, Biotechnology Department, Moscow, 125047, Miusskaya sq. 9,
[3]JSK "Severstal", Cherepovets, Russia

ABSTRACT

With the purpose of improving the removal of nitrogen from the waste water of the coke-chemical plant, a process of biological purification was studied in laboratory conditions and an opportunity of nitrification-denitrification opposite to the classical one was shown. The process takes place in aerobic conditions and without organic substrates. On the synthetic medium with ammonium ions and at the absence of organic substrates it is accompanied by decrease in ammonium concentration without accumulation of nitrite and nitrate in the medium and stimulated by addition of nitrate into the system.

Keywords: ammonia, nitrate, anammox process, oxidation, bioreactors, microbiological degradation, waste water treatment

After removal of phenolics, thiocyanates and cyanides from the waste water of the coke-chemical plant inorganic nitrogen has to be removed at the stage of biological nitrification and denitrification technological scheme of which is shown on fig.1.

* e-mail: aekuz@muctr.edu.ru

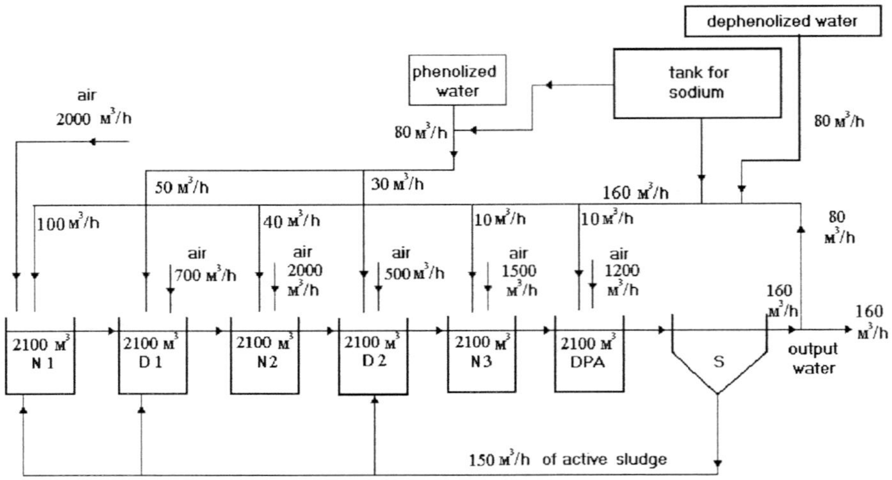

Fig.1. Nitrification-denitrification stage of the waste water treatment of the coke-chemical plant. N – nitrificator, D – denitrificator, DPA – denitrificator-postaerator, S - settler

For this stage of nitrification-denitrification it was found that along with the classical processes of nitrification-denitrification according to the scheme:

$NH_4^+ + O_2 \rightarrow NO_2^- \rightarrow NO_3^-$ $\qquad$ $C_6H_5OH + NO_3^- \rightarrow CO_2 + N_2 + H_2O$

there is a presence of the anammox-like process (Mulder A. et al, 1995):

$NH_4^+ + NO_3^- \rightarrow N_2\ (N_2O) + H_2O$ or $\qquad$ $NH_4^+ + O_2 \rightarrow N_2\ (N_2O)$.

Though we met this process occasionally in 1997 at study of waste water treatment using active sludge with nitrifiers and denitrifiers obtained from the coke-chemical plant of the Cherepovets metallurgical complex, today this seems no surprising taking into consideration i) numerous works of various authors on the anommox process published since 1995 (review see Jetten M.S.M. et al., 2002) and ii) the nitrification-denitrification ecosystem of the coke-chemical plant waste water treatment has suitable ecological conditions to develop such a process.

In our case the process takes place in aerobic condition at pH 6,5-8,5 and without organic substrates. On the synthetic medium with ammonium ions at the range of 100-500 mg/l and at the absence of organic substrates it is accompanied by decrease in ammonium concentration without accumulation of nitrite and nitrate in the medium and stimulated by addition of 20-100 mg/l nitrate ions into the system (fig.2, 5). Addition of NH_4^+ in the medium with NO_3^- without organic substrates stimulates denitrification as well (fig. 3).

The anammox-like process was studied in batch and chemostat conditions and in the bioreactor with filtration (membrane bioreactor, fig. 4). In the bioreactor with filtration the retention of microbial cells and an increase of its capacity were achieved using dynamic membrane formed on the macroporous filter by lignine added into reactor medium. Usage of the principle of dynamic membrane for retention of the biomass allowed to hold the operating capacity of the filter and bioreactor during 3 months of the experiment without regeneration of the filter and renewal of the membrane forming agent (lignine). The rate of NH_4^+ oxidation in continuous conditions for anommox-like process was up to 10 mg/l per h (fig. 5).

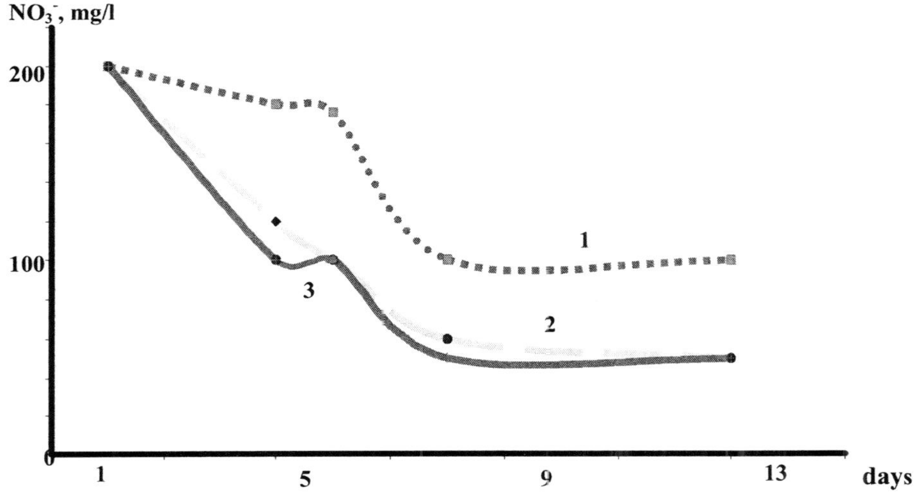

Fig. 2. Influence of NO_3^- -ions on denitrification in aerobic conditions. 1 – 0; 2 – 20; 3 – 40; 4 – 60; 5 – 80; 6 – 100 mg/l of NO_3^-

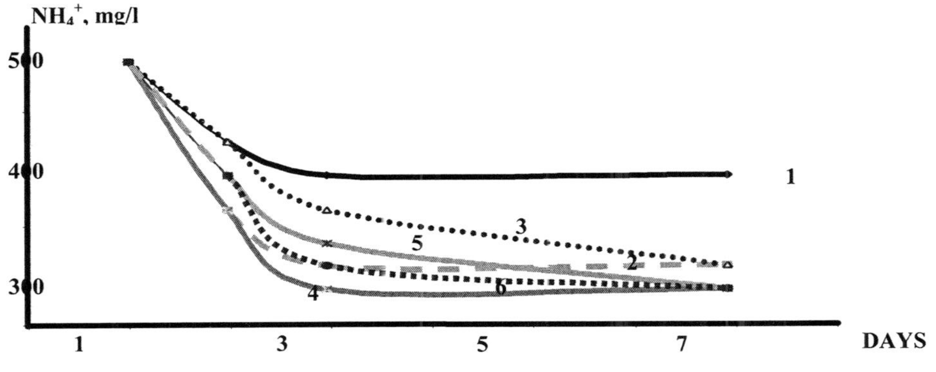

Fig. 3. Influence of NH_4^+-ions on denitrification in anoxic conditions. 1. + 500 mg/l of phenol; 2. + 100 mg/l of NH_4^+; 3 + 60 mg/l of NH_4^+

Influence of the various streams of the coke-chemical plant as well as phenolized and dephenolized waters of the waste water treatment plant on nitrification and denitrification were studied. All the processes and the most probable substrate influences on the nitrogen removal can be described by the following scheme:

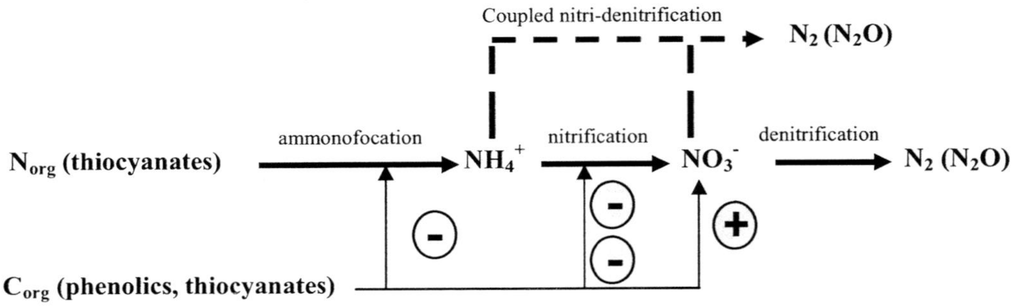

Nitrification is the most susceptible phase to inhibition by phenolics in the process of nitrogen removal from waste water of the coke-chemical plant.

Owing to the concurrent anammox-like process (coupled nitri-denitrification) removal of nitrogen from this waste water requires less amounts of organic substrate initially present or added into waste water. Unbalanced processes of removal of organic contaminants and inorganic nitrogen at the purification as well as lower rate of nitrification-denitrification in contrast with the biodestruction of organic compounds can result in unsatisfactory operation of the waste treatment plant and leads to the increase of the residual amounts of primary and secondary pollutants in the output water stream.

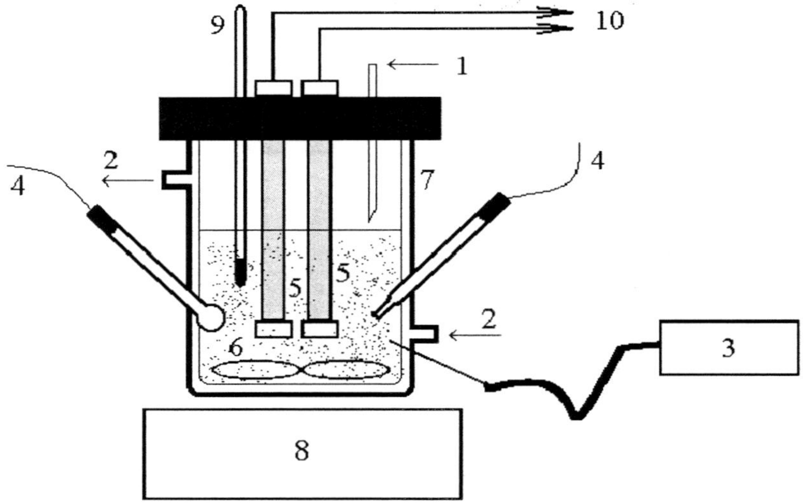

Fig. 4. Bioreactor with dynamic membrane 1 – Inlet of medium, 2 – water for thermostatting, 3 – air pump, 4 – measuring electrodes, 5 – macroporous filter, 6 – suspension of lignine, 7 – reactor shell, 8 – magnetic stirrer, 9 – thermometer, 10 – outlet of filtrate.

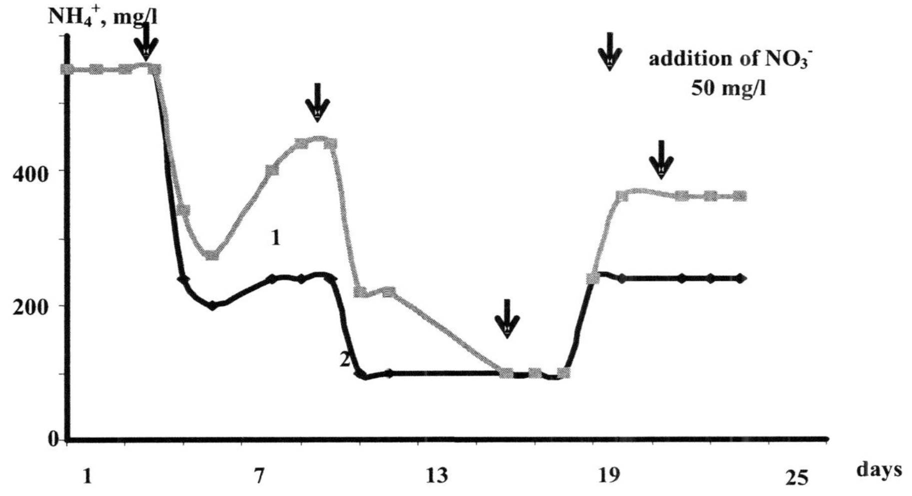

Fig. 5. Influence of nitrate on oxidation of ammonium in continuous culture: 1 – cultivation in chemostat; 2 - cultivation in the bioreactor with dynamic membrane; hours 0-19 – dilution rate 0,2 d^{-1}; hours 19-25 – dilution rate 0,4 d^{-1}. Nitrites were absent in the culture liquid (below 1 mg/l)

REFERENCES

[1] Mulder A., Van de Graaf A.A., Robertson L.A., Kuenen J.G. (1995). Anaerobic ammonium oxidation discovered in a denitrifying fluidized bed reactor. – *FEMS Microbiol. Ecol.,* 16, p. 177-183.

[2] Jetten M.S.M., Schmid M., Schmidt I, et al. (2002). Improved nitrogen removal by application of new nitrogen-cycle bacteria. – *Re/Views in Environmental Science and Bio/Technology,* 1, p. 51-63.

[3] Stepanenkov V.A., Kouznetsov A.Ye., Fu Yigang, Manakov M.N. Normalization of the functioning of the stage of nitrogen removal from the waste water of the JSC "Severstal" coke-chemical plant. – In "Environment for us and future generations: ecology, business and ecological education" – *Proceedings of the Conference "Samara-Astrakhan-Sama*ra", 6-13 of September, 1998, p. 36 (in Russian).

In: New Research on the Environment and Biotechnology ISBN 1-60021-042-2
Editors: A. Kuznetsov and G. Zaikov, pp. 99-104

Chapter 11

Simultaneous Hybrid Processes as a Way to Improve Characteristics of Cultivation of Micro-Organisms and Biodestruction of Pollutants

A. Ye. Kouznetsov*[1] *and S. V. Kalyonov
D.Mendeleyev University of Chemical Technology of Russia
Miusskaya sq., 9, 125047, Moscow, Russia

Abstract

Some systems for investigation of the biosynthesis, biotransformation and biodestruction processes with a simultaneous chemical or photochemical reaction have been developed. It is expected that such hybrid processes will improve biotechnological performances and a variety of the biotechnological methods for pollutant's destruction.

Keywords: hybrid processes, cultivation, oxidation, biodestruction, waste water treatment, bioreactors, artificial peroxisome, hydrogen peroxide, UV-light, photolyase, high density culture.

We define a hybrid process as one where chemical or photochemical reactions and biocatalysis or biodestruction occur in the same volume and simultaneously proceed. They are typical for natural ecosystems. For the purposes of practical application such processes could improve bioavailability of resistant substrate, characteristics of the systems of biopurification and biodestruction, microbial growth and biosynthesis. For scientific research an evolution of these systems in conditions of autoselection has some interest as well.

Examples of various abiotic systems of transformation (oxidative, hydrolytic, reductive) suitable for hybrid processes of cultivation and biodestruction are:

---

[1] e-mail: aekuz@muctr.edu.ru

- Fenton reagent (oxidation of organic substances)

$Fe^{2+} + H_2O_2 \rightarrow Fe^{3+} + OH^- + OH^{\cdot}$

- Kostelfranco reagent (oxidation of organic substances)

ascorbate + Cu^{2+}

- Photochemical oxidation

hv 330-380 nm

$C_{org.} \rightarrow CO_2 + H_2O$

- Advanced photocatalytic oxidation

hv 330-380 nm

$TiO_2 + C_{org.} \rightarrow TiO_2 + CO_2 + H_2O$

- hydrolysis of organic substances, catalyzed by TiO_2, FeOOH, Al_2O_3
- dechlorination by smectites with Cu(II)
- redox processes with oxides Fe (II, III) and Mn (III,IV)etc.

For destruction of resistant chemical pollutants it would be more suitable to use more aggressive abiotic oxidants like hydrogen peroxide or photochemical ones. However there will be a need to solve the problem of maintainance of the phisiological activity of biological component in a hybrid process.

Using model systems with addition of hydrogen peroxide or Fenton reagent (H_2O_2 + Fe^{2+}) into nutrient medium before or during fermentation as well as ultraviolet irradiation (UVA and UVB subrange) of cell suspensions during cultivation several hybrid processes have been studied:

- biodestruction of ion exchange resins as potential waste from atomic power station,
- biodestruction of phenol by an isolate with bacteria and yeast,
- cultivation of yeast *Candida tropicalis* on sucrose,
- cultivation of *Azotobacter* strains,
- biodestruction of phenol by consortia of yeast and bacteria *Azotobacter* on nitrogen free medium.

The summary of the investigations are following:

- pre-treatment of ion exchange resins by Fenton reagent improves their bioavailability;

- phisiological conditions of microbial population for their functioning at the abiotic factors impact, especially being aggressive, are very important for the survival and high activity of micro-organisms;
- at the cultivation in fed batch conditions and at the addition of H_2O_2 the amount of extracellular metabolites and residual concentration of substrate were lower than in the control (without addition of H_2O_2); there was not an accumulation of metabolites inhibiting growth of micro-organisms in the process of phenol biodestruction (fig. 1) and at the growth of *C. tropicalis* on sucrose (fig. 2); these peculiarities kept for a long time after ceasing of H_2O_2 addition;

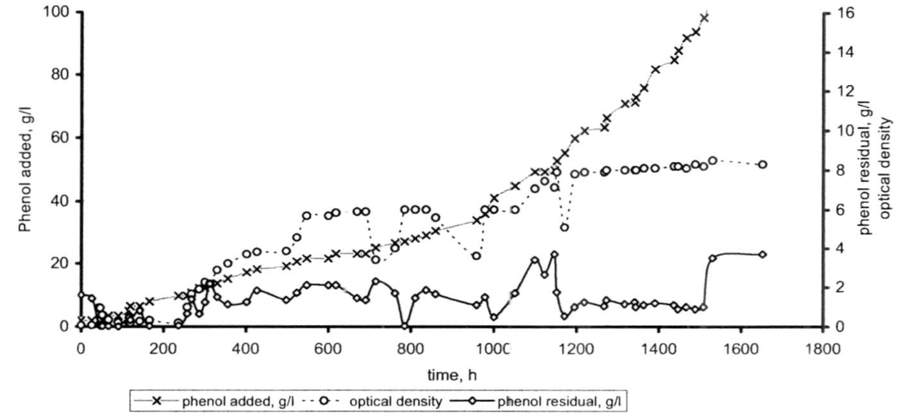

Fig. 1. Biooxidation of phenol in fed batch conditions with fractional addition of phenol and small amounts of H2O2 (since 1 to 1000 h after start of the experiment) into the biooxidation zone and using yeast biocenosis preadapted to hydrogen peroxide.

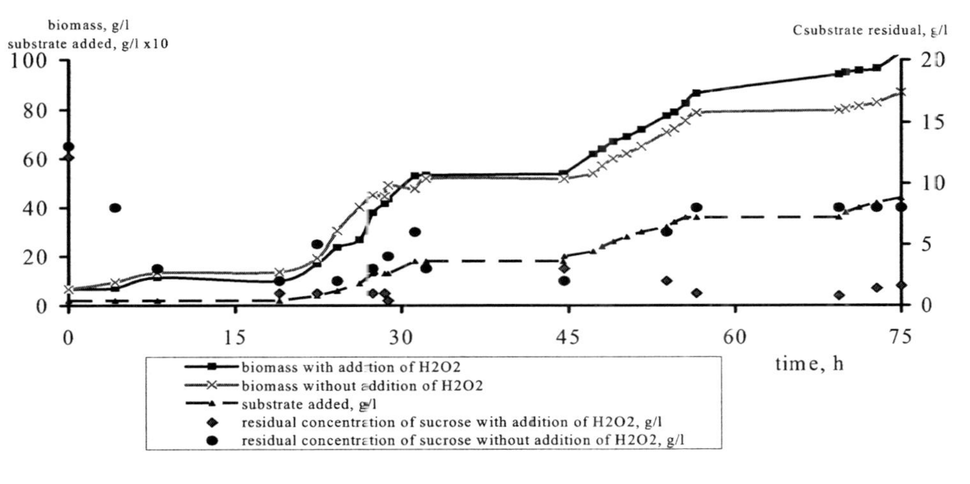

Fig. 2. Growth of yeast *C. tropicalis* on sucrose in fed batch conditions with and without addition of H_2O_2.

- in case of high density cultivation with fractional addition of dry sucrose and periodical addition of H_2O_2 the yeast cells of *C. tropicalis* accumulated up to the maximal level of 150-170 g of dry weight per one liter of culture liquid;
- the rate of oxidation of substrate in fed batch conditions was limited only by oxygen mass-transfer (for biodestruction of phenol and for the growth of *C. tropicalis*);
- diauxy in the consumption of sucrose and extracellular intermediates (acetic acid, ethanol) temporarily accumulated in the culture liquid disappeared (for the growth of *C. tropicalis* on sucrose);
- there was not a dying phisiological activity of cells in fed batch conditions (for biodestruction of phenol and for the growth of *C. tropicalis*);
- cells are more resistant to H_2O_2 at the active stage of the growth (phenol biodestruction, growth of *C. tropicalis*);
- cells pre-adaptated to H_2O_2 were more resistant to various stresses:
  - their activity recovered better after substrate starvation;
  - they could grow at higher running concentration of substrate – up to 5 g of phenol per liter of medium – for phenol biodestruction; up to 700 g of sucrose per liter of medium – for C. tropicalis growth;
  - they were high resistant to hydrogen peroxide: population of phenol biodestructors survived after addition of 2% of H2O2; population of yeast C. tropicalis survived after one-time addition of 16.5% H2O2 (perhydrol diluted 2-fold and burning skin); the last one could be charged by addition up to 15 g of 100% H2O2 per 1 L of medium per 1 hour in continuous condition and could use H2O2 as a source of oxygen;
  - addition of H2O2 to the populations with Azotobacter strains stimulated in some cases their growth on nitrogen free medium with sucrose;
  - there was observed a biodestruction of phenol by consortia of resistant to H2O2 phenol-degrading yeast and bacteria Azotobacter on nitrogen free medium;
  - under simultaneous action of visible and UV-light the yield of yeast C. tropicalis on medium with sucrose was 10-15% above the control (fig 3).

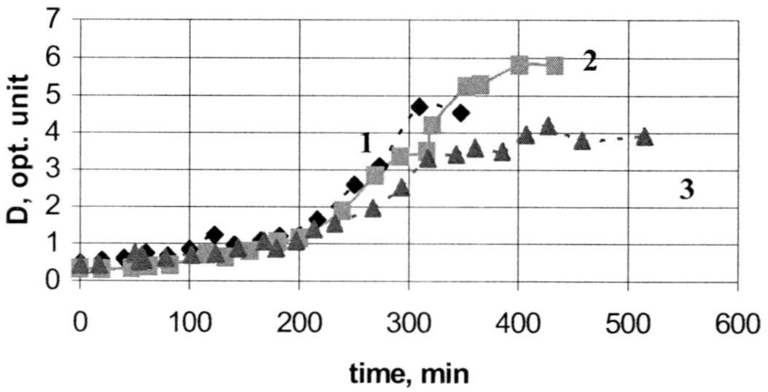

Fig. 3. Growth of yeast *Candida tropicalis* under irradiation of UV- and visible light. 1 – control without UV-irradiation, 2 – with UV- and visible light irradiation, 3 – with UV-irradiation in the dark.

The data obtained in the experiments with simultaneous action of biological and aggressive oxidizing factors (H_2O_2, UV-light) indicate: i) the importance of the supporting optimal red-ox status in the microbial population stipulated, in particular, by concentration of active forms of oxigen; ii) a possibility to achive best output results at the controllable cultivation of micro-organisms at the conditions of active functioning the reparation mechanisms.

For instance, the primary chemical responses initiated by UV-light, H_2O_2 or other forms of active oxygen are accompanied by generation of particles (free radicals etc.), which participate in regulation of physiclogical state of cells, activate anti-stress systems of reparation and initiate the processes of oxidation of biosubstrates.

Among the mechanisms of reparation in cells there is photoreparation by photolyase after UV exposure. The last one functionates together with other ways of response to stress factors. For anti-stress systems of reparation some intersection reactions are possible as well. The system of reparation with photolyase functioning in response to UV irradiation has a maximal sensitivity to light at two wavelength at visible and UV ranges of spectrum. Our preliminary results obtained in the experiments with yeast *Candida* (fig. 3) has shown that one of the ways to improve the efficiency of the hybrid photodestruction and biodestruction processes could be through the simultaneous influence of visible and UV light.

Based on the results obtained three novel methods for biodestruction of pollutants were proposed by us:

- closed intensive microbiological system for the high concentrated waste treatment (fig. 4);
- artificial peroxisome (fig. 5);
- "Solar" photobioreactor (fig. 6).

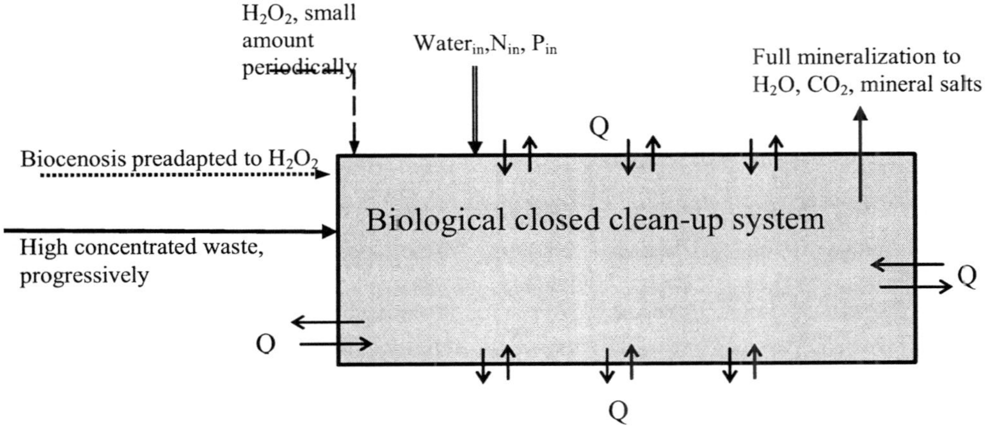

Fig. 4. Closed intensive microbiological system for the high concentrated waste treatment.

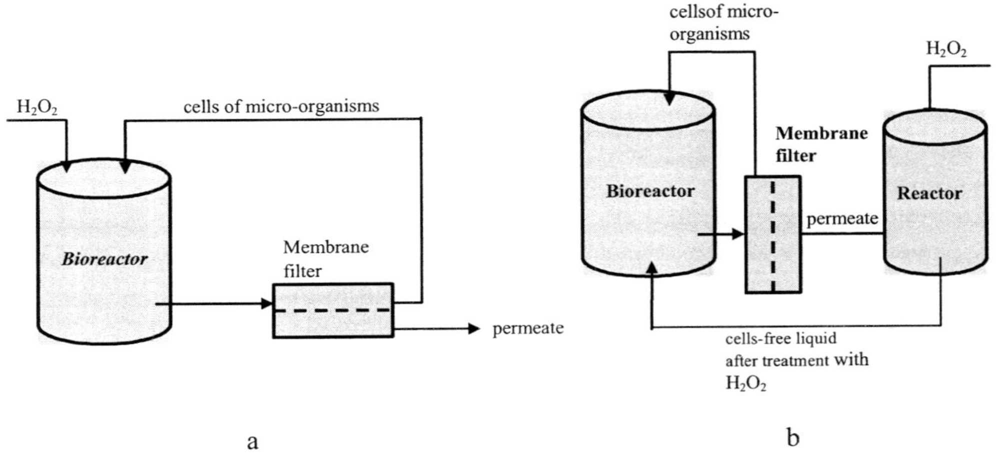

Fig. 5. Artificial peroxisome.

A possibility to protect cells against oxidative stress, induced by addition of H_2O_2 or UV-irradiation, through photoreparation way by photolyase is being under study.

A mathematical model of growth of cells under simultaneous irradiation of visible and UV-light taking into account photoinactivation, photodynamic action, photoreparation and photostimulation will be developed.

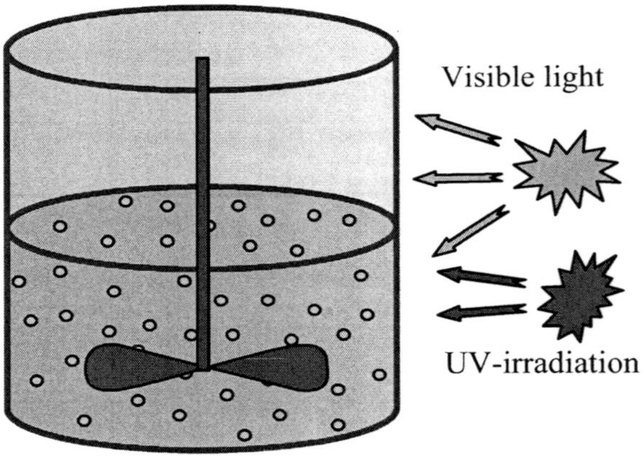

Figure 6. "Solar" photobioreactor.Possible effects: - oxidative stress; - photoreparation; - sinchronisation of cell division; - enhancing substrate bioavailability; - increase of rate of oxidative processes; - high-density culture.

In: New Research on the Environment and Biotechnology ISBN 1-60021-042-2
Editors: A. Kuznetsov and G. Zaikov, pp. 105-113

Chapter 12

BIOREMEDIATION OF SOIL CONTAMINATED WITH ALKYLLEAD COMPOUNDS: LABORATORY INVESTIGATIONS AND FIELD TRIALS

C. Gallert and J. Winter
University of Karlsruhe, Institute for Biology of Engineers and Biotechnology of Wastewater, Am Fasanengarten,
76128 Karlsruhe

Tetraethyllead (TEL) and tetramethyllead (TML) were produced world-wide as anti-knocking additives for gasoline to increase octane numbers. The use of leaded gasoline over almost one century caused an ubiquiteous pollution of the environment with lead ions or alkyllead compounds (Rhue et al. 1992). Nowadays, tertiär-butylether (Europe) or in future ethanol (U.S.) are used to increase the octane numbers.

Whereas car emissions led to an ubiquiteous distribution of low concentrations of organolead compounds and in greater amounts of inorganic lead in the upper layers of soil, spillages of highly toxic tetraalkyllead (TAL) compounds during production, transportation or blending at oil refineries and petrol stations caused more severe soil and groundwater contaminations. The toxicity of alkyllead compounds decreases in the sequence: R_4Pb > R_3Pb^+ > R_2Pb^{2+} with R = $-CH_3$ or $-C_2H_5$. Methylated lead species are less toxic than the ethylated compounds. In the atmosphere TAL is decomposed photocatalytically by UV irradiation, by ozone or by hydroxyl radicals to the more stable, water soluble ionic tri- and dialkylated species. Monoalkyllead cations are assumed to be very instable. After chemical decay and/or biological degradation of organic lead compounds, the inorganic, less toxic lead is immobilized by precipitation as $PbCO_3$ or $Pb_5(PO_4)_3Cl$ in soil. In addition to photodecomposition or oxidative decay, TAL can be decomposed into trialkylated lead species by catalysis of Cu^{2+} or Fe^{2+} ions (Rhue et al. 1992). The ionic trialkyl- and dialkyllead species were stable in an aqueous solution and could not be decomposed by catalysis of Cu^{2+} or Fe^{2+} cations. A sulfide-mediated transalkylation of two moles of trialkyllead compounds into a tetraalkyllead and dialkyllead species was observed by Jarvie et al. (1983). In laboratory experiments and in field studies with soil from an industrial site, that was contaminated with organolead and inorganic mercury, the inorganic mercury was alkylated to ethylmercury by tetraethyllead (Hempel et al. 2000). In addition to photochemical or chemical transformation of alkyllead compounds, microbial degradation in soil occured (Macaskie and Dean 1990, Blais et al. 1990). Biologically mediated multiple dealkylations resulted finally in the formation of inorganic lead.

In Germany TEL and TML were manufactured by two chemical factories until the end of the 1980's. At both production sites significant amounts of organolead compounds were spilled and drained into the underground and still endanger the aquifer. The subsurface soil at the TAL-contaminated site contains mainly fluvial sand deposits with a hydraulic conductivity of 3.5×10^{-4} m s^{-1}. The groundwater table starts at 3 m depth (Gommeringer and Luber 2000). The groundwater flows about 70 m·$year^{-1}$ towards the Rhine river. Upstream it contains about 0.4 mg·l^{-1} oxygen. No oxygen was found in the groundwater downstream of the industrial site. Soil and groundwater samples from bore holes, brought up from the center of the polluted area from 3 - 10 m depth, were highly contaminated with organolead compounds, whereas samples from bore holes in the periphery were less or not contaminated.

For determination of soil respiration activity of contaminated soil, samples M 8-10 (volatile solids content 0.3 %, alkyllead content of the this soil only 0.24 mg·kg^{-1} dry matter) and S 70 (volatile solids content 0.9 %, alkyllead concentration 185.1 mg kg^{-1} dry matter) with and without mineral amendment (N, P, K, Mg fertilizer) were compared in batch assays. Aliquots of 500 g of the soil sample M 8-10 (as a control to determine endogeneous soil respiration) and S 70 (respiration of alkyllead compounds and hydrocarbons) were transfered into 1.2 l glass bottles, closed air-tight and incubated in the dark under aerobic conditions (~21 % O_2). When the oxygen concentration in the bottles had decreased to below 2 %, the gas phase of the bottles was evaccuated and replaced by fresh air.

In both soil samples oxygen consumption and CO_2 production was highly enhanced by mineral addition (Fig. 1a, 1b). In the absence or presence of fertilizer much more oxygen was consumed by the soil containing the high alkyllead and volatile solids concentrations than by the soil, that was only little contaminated with alkyllead and contained less volatile solids (Figure 1a, 1b). The oxygen was, however, not consumed for respiration of alkyllead compounds, but apparently for respiration of oily co-contaminants.

The elution and degradation of organolead compounds from contaminated soil was determined in water-saturated soil columns.

Wet sandy soil from 6 m depth, obtained from drilling in the center of the TAL-contaminated site was homogenized in a technical-size blender (capacity 500 kg). All drain water was adsorbed after intensive mixing for 10 minutes. 18 kg portions of the homogenized soil sample S 80 (90 % dm) were immediately filled into each of 6 glass columns of 10 cm inner diameter and 150 cm length. The glass columns were plugged with rubber stoppers, rapped densely with a silicon tubing for cooling by cold water circulation and with aluminum foil to prevent photooxidation and photosynthesis. Figure 2 shows a scheme of the experimental set-up. For oxygen supply the water in the storage bottle was continuously saturated with pure oxygen to maximally 38 mg/l O_2 and circulated via stainless steel tubings (0.3 mm i.d.) through the column in an upflow mode (soil completely water-saturated). In three of the six columns (S1 – S3) oxygenated water and in the other three columns (S4 – S6) an oxygen-saturated mineral mix (N, P, K, Mg fertilizer) was circulated at a pumping rate of 2.2 l/d, which was about 4 times the flow velocity of the groundwater of ~70 m·$year^{1}$.

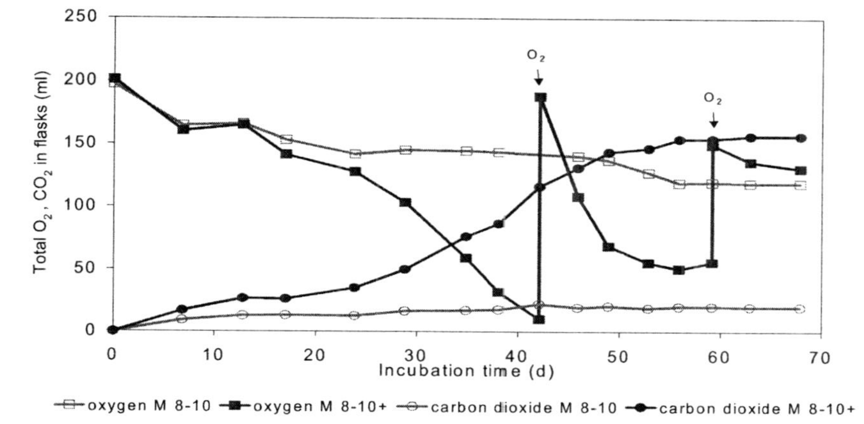

Fig 1a.

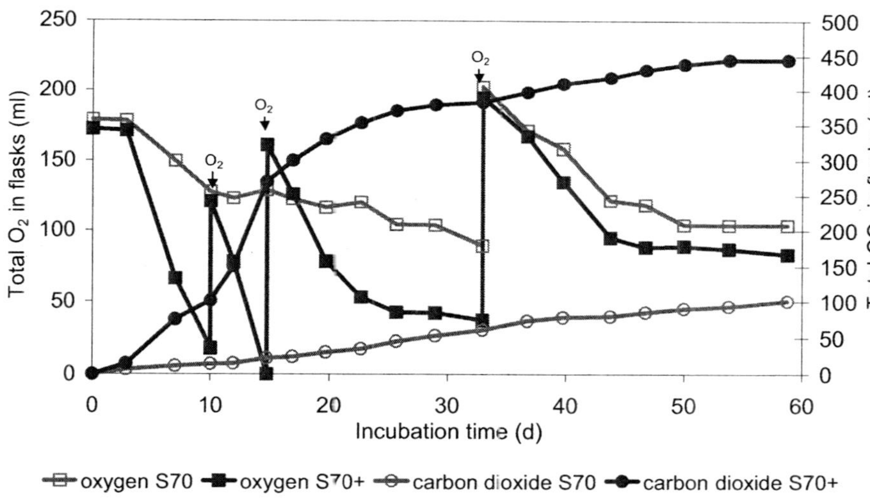

Fig 1b

Fig 1a, b: Oxygen consumption and carbon dioxide production by mixed soil samples M 8-10 (Fig. 1a) or S 70 (Fig. 1b) in the presence or absence of minerals. + indicates fertilization with minerals. Arrows mark an exchange of the headspace of the assays against air.

The average initial content of R_4Pb of the soil samples S 80 was 274 mg kg^{-1} dm, of which 86 % were TEL. The initial concentration of R_3Pb^+ was 22 mg kg^{-1} dm, of which 86 % were triethyllead and the initial concentration of R_2Pb^{2+} was below the detection limit of < 0.5 mg kg^{-1} dm. The mineral hydrocarbon content was 539 mg kg^{-1} dm. No inorganic lead was detected, indicating that no previous mineralization of organolead in the soil samples had occured (Gallert and Winter 2002).

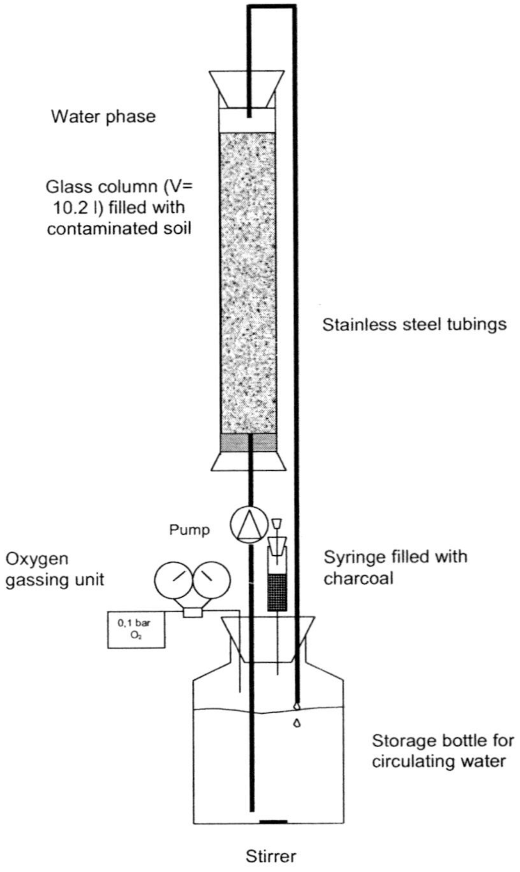

Fig.2: Scheme of the laboratory soil columns used for upflow percolation

A mobilization of TAL compounds from soil into the circulating water and degradation of total alkyllead compounds (R_4Pb, R_3Pb^+ and R_2Pb^{2+}) in the circulating water of column S1 (no fertilizer, Figure 3a) and column S4+ (with fertilizer, Figure 3b) was shown. During the first 98 days of incubation more oxygen was consumed in the fertilized column than in the non-fertilized column, primarily for degradation of non-leaded organic matter. After 220 days of incubation, the maximum of total alkyl lead compounds, mainly trialkyllead compounds was eluted from the soil samples into the circulating water. After that, the concentration slowly decreased with an interim increase of dialkyllead compounds. Although tetraalkyllead compounds were considered insoluble in water, there was some intermediary appearance in the circulating water, presumably due to particle or colloid wash-out (Fig. 3a, 3b. Different scales!).

In order to obtain reliable mass balances of the alkyllead substances during more than 2 years of incubation the experiment was designed as a closed system with water leachate recycling. This led to very high concentrations of e.g. triethyllead, reaching 130 mg l^{-1} after 126 days or even higher than 200 mg l^{-1} after 220 days. During air injection in the field, maximal concentrations of trialkyllead compounds were only 1.2 – 1.5 mg·l^{-1}.

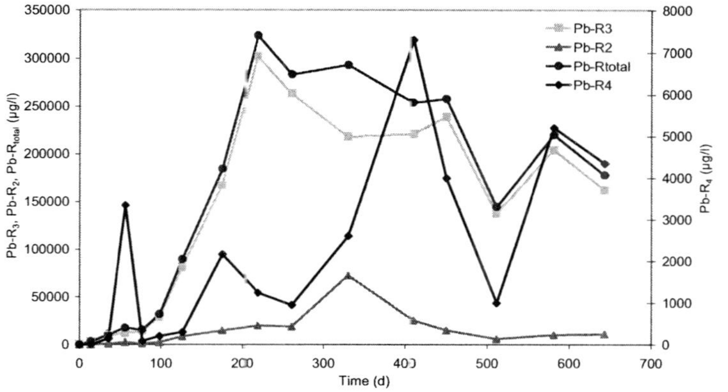

Fig 3a

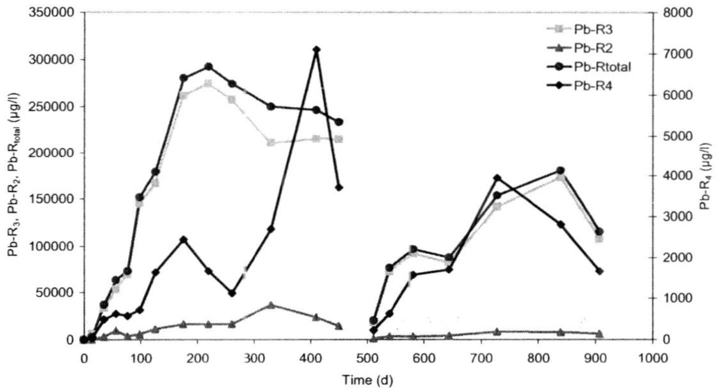

Fig 3b

Fig. 3 a, b: Elution of total alkyllead, tetraalkyllead-, trialkyllead- and dialkyllead-compounds from soil into circulating water and their degradation in the soil columns S1 (unamended) and S4 (amended with a mineral mix).

To examine whether alkyllead concentrations at field and laboratory conditions were toxic, toxicity was tested with *Photobacterium phosphoreum* (DEV, 1983) using a standard solution of triethyllead chloride and as a reference circulating water from one fertilized and one non-fertilized soil column. Triethyllead concentrations causing 25 (EC_{25}), 50 (EC_{50}) or 75 % (EC_{75}) inhibition were 0.2 mg l^{-1}, 1.4 mgl^{-1} or 5.6 mg l^{-1}, respectively. Since the circulating water from soil columns contained much higher triethyllead concentrations, and in addition other trialkyllead-, tetra- and dialkyllead species, inhibition of luminescence could not be related to the triethyllead concentration alone. The inhibiting effect of all of these alkyllead compounds could be reduced to 40 % by a 256-fold dilution of the circulating water. At this high dilution the concentration of triethyllead was only around 0.9 mg l^{-1}. Since the triethyllead concentration in the circulating water was much higher, the soil bacteria must have been much more resistent than the test organism, *Photobacterium phosphoreum*, which represents an unadapted species.

The high concentration and/or the toxicity found in the toxicity test for triethyllead may be the reasons for the slowly decrease of the alkyllead concentrations in the circulating water after reaching the maximum (~ 220 days) in fertilized and non fertilzed soil columns (Fig. 3a, b). An experimental modification after 450 days of incubation should demonstrate the effect of dilution with tap water to non toxic concentrations of triethlyllead concentrations: The circulating water of the 3 non- fertilized soil columns (Fig. 3a) was collected and the soil of each column was flooded three times with tap water. After about 2 weeks, the concentrations of the alkyllead compounds began slowly to increase in the fresh circulating water (Fig. 3a), similar as at the beginning of the experiments. After further 320 days of incubation the maximum of trialkyllead concentrations in the circulating water (not as high as the first maximum) was reached and degradation of alkyllead species was observed. The experiments will be continued until no alkyllead concentrations will be measured in the circulating water.

After 740 days of incubation, extraction and degradation of organolead compounds from contaminated soil in the water-saturated, fertilized soil columns was stopped in order to make a lead balance (Fig. 4). The circulating water was collected and the water-free soil was sampled from the columns. In three different hights the soil was analysed for the occurance of organic and inorganic lead species. The bacterial counts at the beginning and the end of the experiment are shown in Table 1.

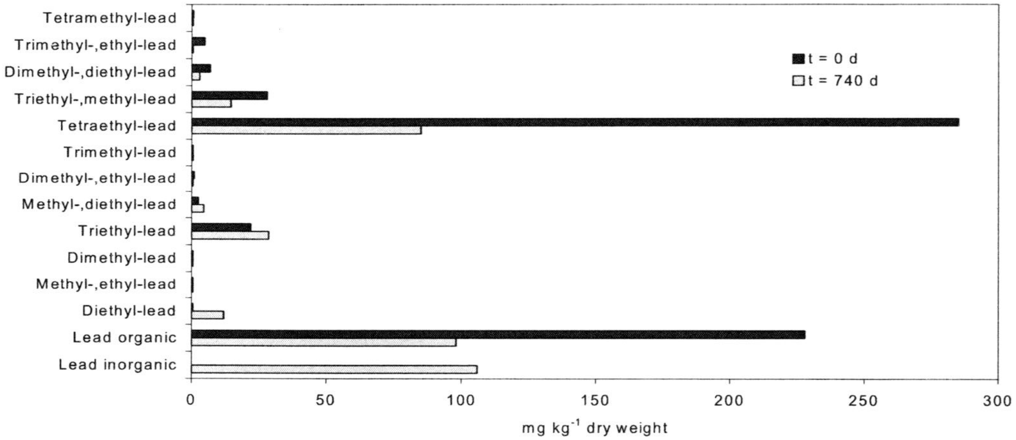

Fig. 4: Mass balance of organic and inorganic lead species of fertilized soil column after 643 days of incubation

Table 1: Determination of bacterial count in fertilized soil columns after 740 days of incubation

CFU (g^{-1} dm)	S4+	S5+	S6+
t = 0 d	7 - 10 x 10^3	1 - 6 x 10^3	1 - 2.4 x 10^4
t = 740 d	5.2 - 8.2 x 10^6	4.7 - 8.2 x 10^6	1.5 - 3 x 10^6

After almost 2 years of incubation, the alkyllead compounds of the fertilized soil were degraded to an extend of 58 %, wereas inorganic lead was formed to an extend of 54 % (Fig. 4). Mineral hydrocarbons were degraded to an extend of 37 %. With these experiments we were able to show the biological degradation of organic lead compounds to inorganic lead during an incubation time of almost 2 years. A negative influence of hydrocarbons on the degradation activity of tetraethyllead was described by Mulroy and Ou (1998). In soil, that was not contaminated with gasoline, TEL was degraded completely during 14 days, whereas in the gasoline-contaminated soil 83 – 96 % of the TEL was degraded only after 77 days. In TAL-contaminated soil columns transformation of tetraalkyllead into tri- and dialkyllead was slow and was not significantly influenced by a concomittant hydrocarbon degradation in the presence or absence of minerals.

The laboratory experiments were created as a "closed system" in order to make an accurate mass balance. The negative effects were the high and toxic concentrations of alkyllead compounds in the circulating water. The soil columns with diluted circulating water (Fig. 3a) were incubated for a longer time in order to degrade most of the organic alkyllead species to inorganic lead.

To simulate the conditions in the field which means concentrations of alkyllead species in the groundwater of 1-10 mg l^{-1}, a new set of soil columns was started. Non contaminated sand from the Rhine valley was filled into 3 columns and diluted circulating water of one of the contaminated laboratory columns was used. The time dependence of degradation of alkyllead species is shown in Fig. 5.

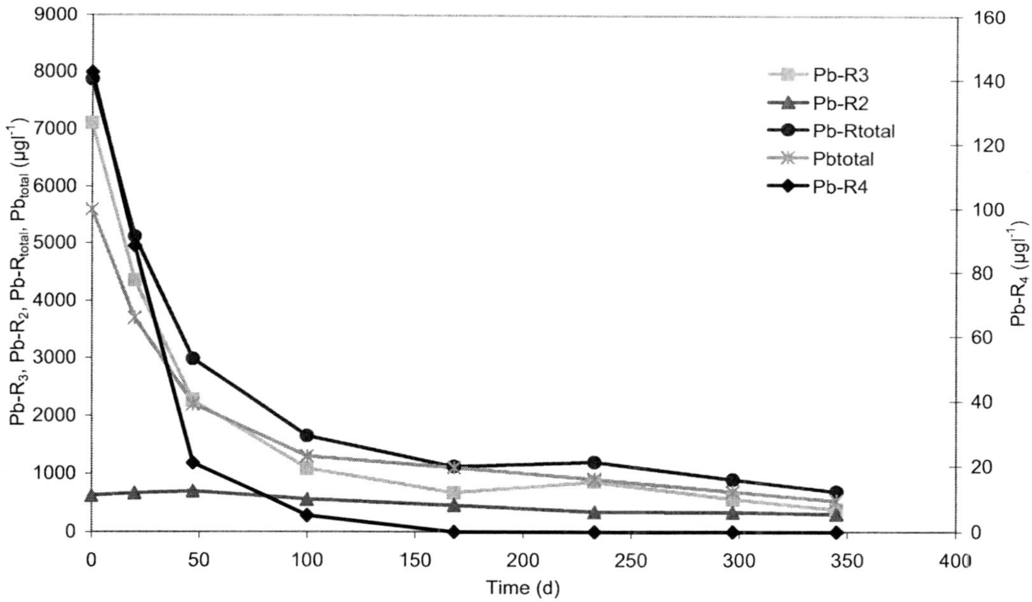

Fig. 5: Degradation of alkyllead compounds from ground water: application of field concentrations

If the initial concentrations did not exeed 8 mgl^{-1}, after 168 days of incubation, almost 100 % of tetraalkyllead compounds and after 345 days 95 % of trialkyllead compounds were degraded. The dialkyllead compounds, the degradation product of tetra- and trialkyllead species were degraded to an extend of 51 %. The time needed for degradation may be reduced by addition of an alkyllead degrading population or by feeding the almost "sterile" sand with more alkyllead contaminated water to enhance the biological activity.

Microorganisms that were capable of degrading tetraalkyllead have not jet been isolated, presumably due to the low water solubility of this substrate of 6 – 9 x 10^{-7} mol l^{-1} (Rhue et al.

1992). Macaskie and Dean (1990) worked with an *Arthrobacter* spec., which was able to grow in a medium that contained 100 μg Pb · l^{-1} as trimethyllead.

To increase the oxygen concentration of the groundwater, air injection wells (AIW) were installed at the contaminated industrial site. The well itself consists of a perforated metal tube reaching 4 m into the groundwater table (Fig. 5). A smaller air injection pipe is fitted into the perforated pipe to inject air from the surface. The air is released at the base of the tube and due to the buoyancy the air moves upwards lifting the water level in the perforated pipe above the ground water level. The result is a groundwater circulation aerea in the aquifer (Luber 1999) and an enrichment of the groundwater with oxygen or other supplements. In the first phase, 4 AIW with a distance of 5 m were installed. Meassurement of the oxygen concentration in monitoring wells showed an increase of oxygen of one order of magnitude during the first weeks of operation. Parallel with the enrichment of oxygen, the tri- and dialkyllead concentration in the groundwater increased due to a higher avilalability of oxygen.

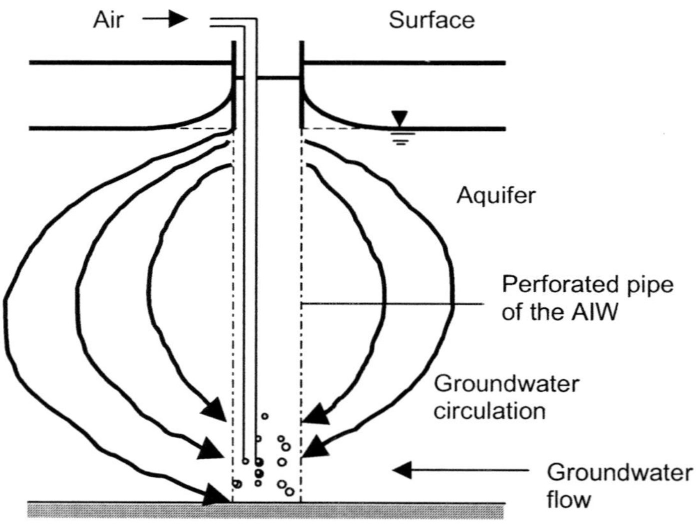

Fig. 6: Air injection well AIW and groundwater circulation during air injection according to Luber (1999)

Analyses of alkyllead compounds in the soil and groundwater at the industrial site before and after air injection led to the conclusion, that in the natural environment not sufficient oxygen is available for a chemical and/or biological transformation of tetraalkyllead compounds to tri- and dialkyllead species. Whether the in situ distance of the hot spot from the boundary of the property of about 150 m in the direction of the groundwater flow is sufficient for a complete degradation of water-soluble trialkyllead via dialkyl to inorganic lead, if enough oxygen is supplied, remains to be determined. From our laboratory experiments it can be concluded, that a bioremediation of alkyllead contaminants in principle is possible, if enough oxygen is supplied. Alternatively, nitrate or sulfate might be suitable electron acceptors.

REFERENCES

Blais, J. S., Doige, C. A., Marshall, W. D. and Knowles, R. (1990). Persistence and toxicity of alkyllead salts to anaerobic nitrogen transformations in soil. *Arch. Environ. Contam. Toxicol.*, **19**: 227-234.

Gallert C., Winter, J. (2002) Bioremediation of Soil Contaminated with Alkyllead Compounds. *Wat Res* **36**: 3130-3140.

Gommeringer, K. and Luber, M. (2000). In situ groundwater remediation using air injection wells and the principle of groundwater circulation. *International workshop Contaminated Sites Assesment and Remediation,* 10. Nov. 2000, Mailand.

Hempel, M., Kuballa, J. and Jantzen, E. (2000). Discovery of a transalkylation mechanism – Identification of ethylmercury+ at a tetraethyllead-contaminated site using sodiumtetrapropylborate, GC-AED and HPLC-AFS. *Fresenius J. Anal. Chem.*, **366**: 470-475.

Jarvie, A. W. P., Withmore, A. P., Markall, R. N. and Potter, H. R. (1983). The sulfide assisted decomposition of trimethyllead chloride in aqueous systems. *Environ. Pollut. Ser. B. Chem. Phys.*, **6**: 69-79.

Luber, M. (1999). Luftinjektionsbrunnen zur in situ Grundwassersanierung – Hydraulik und Stoffdurchgang. *Doctoral Thesis* at the Institute of Soil Mechanics and Rock Mechanics, University of Karlsruhe.

Macaskie, L. E. and Dean, A. C. R. (1990). Trimethyl lead degradation by free and immobilized cells of an *Arthrobacter* sp. and by the wood decay fungus *Phaelus schweinitzii*. *Appl. Microbiol. Biotechnol.*, **33**: 81-87.

Mulroy, P. and Ou, L.-T. (1998). Degradation of tetraethyllead during the degradation of leaded gasoline hydrocarbons in soil. *Environ. Toxicol. Chem.*, **5**: 777-782.

Rhue, R. D., Mansell, R. S., Ou, L.-T., Cox, R., Tang, S. R. and Ouyang, Y. (1992). The fate and behaviour of lead alkyls in the environment: A review. *Crit. Rev. Environ. Control*, **22**: 169-193.

In: New Research on the Environment and Biotechnology ISBN 1-60021-042-2
Editors: A. Kuznetsov and G. Zaikov, pp. 115-121

Chapter 13

TWO-STAGE RECYCLING TECHNOLOGY OF NON-FERROUS METAL'S BACTERIAL LEACHING FROM SULFIDE RAW MATERIALS

N. Fomchenko, O. Slavkina[1], V. Biryukov, M. Arkhipov and A. Kuznetsov

Moscow State University of Ecological Engineering.
107066, Moscow, Russia

ABSTRACT

Two-stage recycling technology of non-ferrous metal's sulfide product bacterial leaching was developed on the base of sulfide copper-zinc flotation product of Uchalinsky ore mining and processing enterprise. The influence of different technological parameters on the efficiency of sulfide product leaching was investigated for both stages. The laboratory unit for continuous two-stage leaching of sulfide copper-zinc product was developed and tested. The principal possibility of the more complete zinc and copper recovery was shown in comparison with single-stage leaching process. The trials showed that in definite conditions during 24-hours of leaching it is possible to extract 91-93% of zinc, while its content in leached material is decreased to 1,7-1,3 %.

Key words: non-ferrous metals, bacterial leaching, ferric iron, ferrous iron, sulfide product, Acidithiobacillus ferrooxidans.

INTRODUCTION

It is well known that zinc ores of Russian deposits are extremely difficult for concentration. A great deal of nonstandard middling and concentrates are generated during

[1] Address for correspondence: Staraya Basmannaya st., 21/4.,Tel (fax): (095) 267-07-28,E-mail: slavkinaOlga@rambler.ru

their processing. Today such products could be processed only by means of combined technologies involving bacterial leaching, because using traditional methods is not economically and ecologically appropriate and rather complicated [1-3].

However, the bacterial leaching of non-ferrous metal's sulfide products is characterized by low rates of target components transferring into the solution. That's why it needs intensification [4, 5].

It is worthwhile to subdivide the process of metal's bacterial leaching into two stages for its intensification. The process of sulfide's oxidation and metal's leaching under the action of ferric iron must be realized in the first stage. Here the ferric iron is reducing to ferrous iron and transporting to the second stage together with a thin-dispersed fraction of leaching material. The oxidizer regeneration and thin-dispersed fraction complete leaching takes place in the second stage under the active participation of bacteria *Acidithiobacillus ferrooxidans*.

According to this approach it is possible to provide optimal conditions for the proceeding of each stage.

In this report we would like to present the results of studying the first and second stages of sulfide zinc-copper product leaching and also the results of two stage recycling technology trials.

Materials and Methods

All investigations were carried out with zinc-copper sulfide product, containing 10,0-16,7% Zn, 0.9-1.5% Cu, and also 34,5-45,0 % S and Fe, obtained at Uchalinsky ore mining and processing enterprise. The approximate product content was (%): blende (18,5 – 21,0), chalcopyrite (1,6 - 2,1), pyrite (64,0 - 71,0). In all experiments the quantity of liquid phase (bacterial solution) was 100 ml while the solid phase quantity was taken as 5, 10, 15, 25, 50 and 70 g (4.8, 9.1, 13.0, 20.0, 33.3, 41.2 % correspondingly).

The previously prepared cultural broth of bacteria *A. ferrooxidans* (kindly granted to us by G.I. Karavayko, Institute of Microbiology of Russian Academy of Science) grown at the standard medium 9K was used as a leaching solutions, where different concentrations of ferric iron were achieved by means of initial ferrous iron oxidation by bacteria.

During experiments there were measured the concentrations of $Fe^{2+}$and Fe^{3+} (chelatometric titration with trilon B [6]), Cu^{2+}, Zn^{2+} in liquid and solid phases (atomic absorption) and also pH (potentiometry). The contents of general sulphur, sulphate ions, and sums of pyrite and sulfide sulphur were measured with the help of phase-analytical methods. The oxidation-reduction processes effectiveness was considered by the studied parameters time history. Target components leaching were estimated as rate of zinc and copper accumulation in liquid phase and rate of non-ferrous metals content decreasing in solid phase [5].

The proposed recycling technology was tested at specially mounted pilot laboratory unit containing bioreactor with volume of 0,68 l on the base of "Bioflo" fermenter (New Brunswick, USA) with modernized agitation system, sedimentation tank for suspension separation, bubble fermenter with temperature control and volume of 32 l, peristaltic pumps for streams and recycle organization, thermostats of continuous operation and sedimentation tank for thin-dispersed fraction after the bubble fermenter.

RESULTS AND DISCUSSION

The influence of different technological parameters on the both stages of leaching was studied in flasks in periodical conditions.

The First Stage of Leaching

The process intensity was shown to rise with solid phase mass concentration (suspension density) increasing within the range from 4,8 to 41.2%, but after 33.3% this increasing stops. Hence, the excessive solid phase concentration increasing aggravates the conditions for leaching process itself. The leaching process carrying out was also studied at three different ferric iron concentrations (5, 10 and 20 g/l). The ferric iron concentration increasing from 5 to 10 g/l leads to considerable leaching rate rising (about 30 %), while the iron concentration increasing up to 20 g/l leads to comparatively small rising of leaching rate (less than 10 %). The further ferric iron concentration increasing is also not advisable because of jarosite precipitation and leaching conditions worsening.

Fe^{2+} concentration increasing was found to result in rate of bioleaching decreasing. And this fact can partially explain not proportional growth of process rate with ferric iron concentration rising from 5 to 10 and from 10 to 20 g/l (because in the latter case the liquid phase has more quantity of ferrous iron) [6].

It was stated that the presence of bacteria *A. ferrooxidans* in leaching solution is not effect the process intensity at least at suspension density 33 %. It seems to be connected with high solid phase concentration and as the result the high rate of ferric iron reduction to ferrous iron. In addition bacteria loose their livability at high solid phase content. No difference was observed in the rates of leaching at leaching solution initial pH value varying within the limits of 1.4-1.8.

Zinc ions was shown to have no inhibiting influence at leaching process (at zinc concentrations below 50 g/l) and on the contrary the considerable inhibiting of leaching process was observed for copper ions already at the copper concentration of 5 g/l.

The temperature was found to be on of the determining factors that mainly affect the leaching process rate. The initial process rate (at the moment of 1 hour) is 5,6 times more at temperature of 65°C than at room temperature (20 ^{0}C).

The influence of agitation intensity is quite obvious and at given stage it is not connected with dissolved oxygen presence or absence at the medium [7].

The experiments conducted permitted to formulate the requirements for intensive leaching process carrying out at the first stage: Fe^{3+} concentration - 10-12 g/l, solid phase concentration 33,3 %, pH – not higher than 1,6, temperature 65 ^{0}C. It should be noted that optimal temperature meaning at the first stage will be determined by process hardware implementation.

The Second Stage of Leaching

The aim of this stage is to obtain Fe^{3+} and SO_4^{2-} ions being necessary for the first stage (SO_4^{2-} - ions take part in ferric iron regeneration according to reaction $4FeSO_4 + O_2 + 2H_2SO_4 \rightarrow 2Fe_2(SO_4)_3 + 2H_2O$, while the formation of sulphate-ions itself is going according to the reaction: $2S^0 + 3O_2 + 2H_2O \rightarrow 2H_2SO_4$.) Solution with ferrous iron and thin-dispersed fraction of leaching material is constantly occuring at the second stage at continuous leaching process organization. Ferrous iron oxidation together with necessary sulphate - ions obtaining accompanied by complete non-ferrous metals leaching are realized with the help of bacteria *A. ferrooxidans*. Laboratory investigations showed that at solid phase concentration 1-5 % the ferric ions regeneration process intensity is mainly determined by activity of the bacterial culture used. The rate of ferrous iron oxidation achieved was 2 g/(l×hour) at obtaining leaching tails with the following residual concentrations: zinc - 1,05 %, copper - 0,31 % and total recovery of zinc - 93,6 %, and copper – 80,6 % [8].

Two Stage Recycling Technology of Zinc-Copper Sulfide Product Bacterial Leaching

The obtained laboratory results permitted to propose the technological scheme of zinc-copper product proceeding (Figure 1)

The conditions of intensive agitation at temperature of 65-75 ^{0}C and solid phase concentration of 30-40 % are maintained at the first stage in reactor 1. The regenerated solution of ferric iron with concentration of 10-12 g/l as the oxidizing agent is constantly transferring into reactor 1 at simultaneous removing of reduced iron Fe^{2+} and leached non-ferrous metals Cu^{2+} and Zn^{2+}containing solution. The suspended thick-dispersed fraction is separated from the stream in sedimentation tank or separator 2 and continuously returning to the reactor of the first stage [8-9].

Then the leaching solution with small quantity of thin-dispersed fraction (~1%) goes to the second stage in bioreactor 3 where kept conditions being optimal for microorganisms *A. ferrooxidans* vital activity (temperature 0-32 ^{0}C, pH 1,6 – 1,9, intensive aeration). Here happens the ferrous iron oxidation to ferric iron and it allows to send regenerated solution back to the reactor 1 of the first stage.

A part of the stream from sedimentation tank 2 is periodically transported to the unit of non-ferrous metals isolation 5. This solution actually contains no ferric iron, whose presence affect adversely on the process of zinc and copper recovery. After the unit of zinc and copper isolation the spent solution returns to bioreactor 3, thus taking part in closed process liquid cycle.

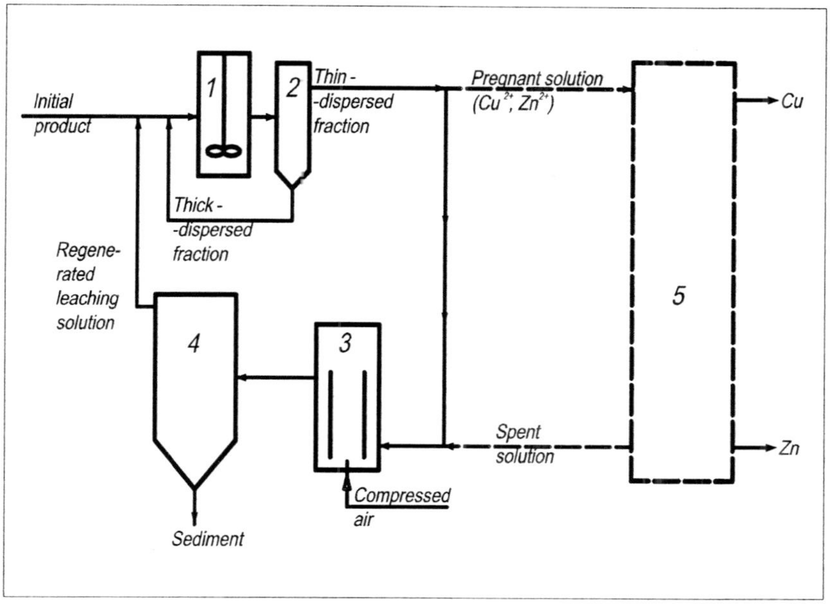

Figure 1. The scheme of zinc-copper pyrite product proceeding with the help of two stage recycling technology.1. - Reactor, 2,4. - Sedimentation tanks, 3.– Bioreacotor, 5. – Unit of non-ferrous metals isolation.

The process control system suppose the correction of solution recycling rate into the reactor 1 according to ferrous iron concentration in bioreactor 3 and also the maintaining of intensive aeration in bioreactor 3 in dependence from dissolved oxygen concentration within the range of 10-15 % from saturation [7]. It permits to coordinate the rate of solution recycling to the first stage with bacterial culture activity, providing the ferric iron regeneration in bioreactor.

The technological scheme proposed is characterized by closed liquid streams cycle. Its waste is non-toxic leached product with zinc concentration 1,7 % and even lower. The gases pollutants take place only after the bioreactor but they consist no toxic substances. (Although there is the possibility of sulphuric acid vapors exhausting from the bioreactor, they are easy in recovery.)

Process Testing

To test the described technology of two-stage bacterial leaching of Uchalinsky product, the special laboratory pilot unit was mounted.

The trials were conducted with total solid phase returning from the sedimentation tank to reactor of the first stage. The initial concentration of solid phase (with zinc and copper content 15,3 and 1,5 % correspondingly) in the reactor was 500 g/l.

The advantageous conditions for the first stage of leaching maintained in the reactor 1 were: temperature 65 ^{0}C, pH – 1,6-1,8, intensive agitation, inlet ferric iron concentration 10-12 g/l. Optimal conditions for ferric iron regeneration with the help of bacteria maintained at the second stage of bacterial leaching in bioreactor were: temperature 28-30 ^{0}C, pH – 1,6-1,8, aeration 1 l/(l×min).

The influence of the following parameters on the leaching process efficiency was investigated: flow rate through the system in the range of 1,5 – 4,6 volume/(volume×hour), the rate of solid phase recycling from the sedimentation tank to the first stage, temperature and pH. The rate of ferric iron reduction at the first stage, rate of ferrous iron oxidation by bacteria at the second stage and also pH were measured during trials.

The best results were obtained at flow rate 2 l/h (3 volume/(volume×hour)), at average ferric iron reduction rate 8,8 g/l×h and ferrous iron oxidation at the second stage 1 g/l×h, when the zinc and copper leaching rates of 2,01 and 0,17 g/l×h were achieved. Table 1 represents the results of leaching residuals analysis.

As it is follows from the above mentioned data, the two stage technology realization permitted to recover 93 % of zinc and 58,9 % of copper after 24 hours process already at the first stage at total sulfide sulphur oxidation of 67 %. After the second stage zinc and copper recovery is increased up to 94,6 and 82,9 % correspondingly.

The technological characteristics and activities found out are acceptable for industrial production organization. In comparison with traditional one-stage method, the process duration was decreased from 110 h [2] down to 24 h, while the residual zinc content in residuals was also reduced from 3,5 – 4 % to 1,3 %.

For further technological scheme optimization it is necessary to study the following questions:

1. Possibility to increase the productivity of the first stage with the help of continuous solid phase supply into the reactor 1 (keeping constant high concentration of solid phase in reactor 1 and inflow of initial product with unoxidized surface provide the high rate of product oxidation)
2. Opportunity to use the thermophilic microorganisms, capable to active sulfides and ferrous iron oxidation at temperature of 60 °C at the second stage.

Table 1. The main results of trials of two-stage recycling technology of zinc and copper bacterial leaching from Uchalinsky sulfide product. (Solid phase concentration 33,3%, T-65°C, time–24h).

№	Products	Content, %			Recovery, %		
		Zn	Cu	S (sulfide form)	Zn	Cu	S (sulfide form)
1	Initial product	15,25	1,5	48,8	100	100	100
2	Solid phase after the first stage (8 hours)	6,13	1,3	42,6	60	13,3	12,7
3	Solid phase after the first stage (24 hours)	1,33	0,77	20, 15	93,0	58,9	67,0
4	Solid phase after the second stage (24 hours)	1,03	0,34	11,23	94,6	82,9	81,6

REFERENCES

[1] G.I. Karavaiko: Microbiological processes of metals leaching from the ores. Review. *Metals biogeotechnology*. CMP GKNT. Moscow, 1984. 87 p. (in Russian).

[2] V.V. Panin, A.V. Adamov et al.: The using of bacterial leaching technology at multiple copper-zinc ores dressing. *Zvetnye metally*, 5 (3), (1999). (in Russian).

[3] J.A. Brierly, C.L. Brierly: Present and future commercial applications of biohydrometallurgy. *Hydrometallurgy*, 94, 99 (2001).

[4] Y. Konishi, H. Kubo, S. Asai: Bioleaching of Zinc Sulfide Concentrate by Thiobacillus ferrooxidans. *Biotech. Bioeng*, 39, 566 (1992).

[5] N. Fomchenko, O. Slavkina, V. Biryukov: Biological oxidation of sulfide raw material using a culture of *Thiobacillus ferrooxidans* under various conditions of leaching. *Applied biochemistry and microbiology*, 39, 82 (2003). (in Russian).

[6] N. Fomchenko, O. Slavkina, V. Biryukov: The studying of copper-zinc ore concentrate bacterial leaching. I. The investigation of different factors influence on the leaching rate. *Biotechnology*, 4, 80 (2002). (in Russian).

[7] N. Fomchenko, O. Slavkina, V. Biryukov: Ecologically safe technology of non-ferrous metals obtaining from sulfide raw materials by means of bacterial leaching. *Yasawa international symposium on metallurgical and materials processing: principles and technologies. Aqueous and Electrochemical Processing.* The United States of America, 3, 337 (2003).

[8] N. Fomchenko, O. Slavkina, V. Biryukov: The development of new technology for non-ferrous metals obtaining from sulfide raw materials: influence of different parameters on the kinetics of primary bacterial leaching. *6-th conference on environment and mineral processing.* Czech Republic, 1, 405 (2002).

[9] N. SMALLEY, G. DAVIS: Operation of copper sulphides concentrate using extreme thermophilic bacteria. *Minerals Engineering*, 8 (12), 56 (1999).

In: New Research on the Environment and Biotechnology ISBN 1-60021-042-2
Editors: A. Kuznetsov and G. Zaikov, pp. 123-133

Chapter 14

DESTRUCTION OF CYANIDES AND THIOCYANATES

N. V. Grigor'eva[1],*, T. F. Kondrat'eva*, E. E. Savari**, G. V. Sedel'nikova** and G. I. Karavaiko*

* Russian Academy of Sciences Winogradsky Institute of Microbiology, Prospekt 60-let Oktyabrya, 7/2, 117312 , Moscow, Russia

** Central Research Institute of Geological Prospecting for Base and Precious Metals (TsNIGRI), Varshavskoe sh. 129 b, 117545 Moscow, Russia.

ABSTRACT

The screening of microorganisms resulted in the selection of a bacterial consortium, *Pseudomonas putida*, strain 21, and *Pseudomonas stutzeri*, strain 18, which is active in the destruction of CN^- and SCN^-. The mechanism of CN^- and SCN^- destruction by microorganisms was elucidated. In the presence of an organic substrate these compounds are utilized as a nitrogen source. CN^- destruction yields NH_4^+, whereas SCN^- destruction produces cyanate, which is then transformed into NH_3. Sulphur is oxidized to thiosulphate and polythionates.

A method of bacterial-chemical destruction of cyanides and thiocyanates in industrial wastewaters and pulps is developed, which makes it possible to eliminate their high concentrations (200 mg/l CN^-, 4-6 g/l SCN^-). The method involves two stages, chemical and microbiological. The chemical processing of wastewaters uses a metabisulphite of an alkaline metal in the presence of copper sulphate, as a result of which cyanide is transformed into a less toxic cyanate and then into ammonium. Thiocyanates and residual cyanides are destroyed using the aforementioned consortium of heterotrophic bacteria in the presence of phosphates and ethanol as an organic source. The wastewaters were purified from CN^- and SCN^- down to the maximum permissible concentration of 0.05 mg/l.

[1] E-mail: gregor@inmi.host.ru,Fax/Phone (095) 135-65-30

Keywords: cyanides, thiocyanates, consortium, heterotrophic bacteria, bacterial-chemical technology.

1. Introduction

Cyanides are used in the extraction of precious metals from ores, in galvanic gold and silver plating, in photography, and in organic synthesis. Up to 3 million tons of cyanides find industrial use every year. These are highly toxic compounds. Thiocyanates are produced mainly from cyanides upon reaction with polysulphides and sulphide minerals when gold is being extracted from ores and sulphide concentrates. Detoxification of industrial wastewaters containing cyanides and thiocyanates is a very important problem. There are several ways to neutralise industrial wastewaters: chemical, microbiological and combined methods. The chemical approach is the main one, with alkaline chlorination and oxidation with sulphurous anhydride in the presence of a copper catalyst (the INCO technique) being the two most widely employed.

Both methods have substantial shortcomings: alkaline chlorination leaves high concentrations of chlorine (up to 300 mg/l) and chlorine organic compounds, whereas the INCO technique is not effective in reducing cyanide content to the maximum permissible concentration (0.05 mg/l) and entirely ineffective in thiocyanate detoxification. Other chemical methods involve the precipitation of water-insoluble iron cyanide complexes, hydrolysis at high temperatures, treatment with acids (sublimation of cyanides) and adsorption. However, their application also has drawbacks, such as the high cost of equipment, the corrosive effect of the medium, etc.

Microbiological methods of detoxification of wastewaters containing cyanides and thiocyanates are effective, economical and ecologically safe.

There is a broad range of microorganisms capable of decomposing cyanides including (a) bacteria such as *Pseudomonas sp., P. fluorescens, P.putida, P. aeruginosa, P.stutzeri, P. paucimobilis, P. pseudoalcaligens*; *Escherichia coli, Alcaligenes xylosooxidans denitrificans, Arthrobacter sp., Acinetobacter sp., Thiobacillus denitrificans, Klebsiella ozaenae, K. oxytoca, Bacillus pumilus, B. stearothermophilus, B. megaterium, B. subtilis, Nocardia sp.* and (b) fungi including *Stemphylium loti, Gliocladium virens, Trichoderma koningii, Fusarium oxysporum, F.solanii, F.lateritium, Aspergillus niger, Gloeocercospora sorghi, Neurospora crassa, Leptosphaeria makulans.* Far fewer microorganisms capable of thiocyanate degradation are known: *Thiobacillus thioparus, Paracoccus thiocyanatus, Thioalkalovibrio sp., Pseudomonas stutzeri, P. fluorescens, Arthrobacter sp., Netsieria meningitides.*

The first industrial installation for biodegradation of cyanides and thiocyanates was built in the USA (the Homestake plant for the extraction of non-ferrous and precious metals) (Whitlock J.L., 1990). Wastewaters containing up to 11.5 mg/l of cyanides and up to 110 mg/l of thiocyanates, i.e. relatively low concentrations of these compounds, were detoxified, using a mutant strain of *Pseudomonas paucimobilis* and nitrifying bacteria in the composition of activated sludge. When the same method was applied in order to detoxify high (100 mg/l) CN^- concentrations, the degree of cyanide destruction was as low as ca. 43 per cent.

The main factors limiting the efficacy of microbiological methods are cyanide concentration in the medium and pH. High (> 100 mg/l) cyanide concentrations are toxic for the majority of microorganisms. Within the pH range which is most favourable for microbial growth (pH 7-8), cyanide is present in the state of weakly dissociated acid (HCN) and its loss is possible in the course of destruction (transition into the aerial phase) (Fig. 1).

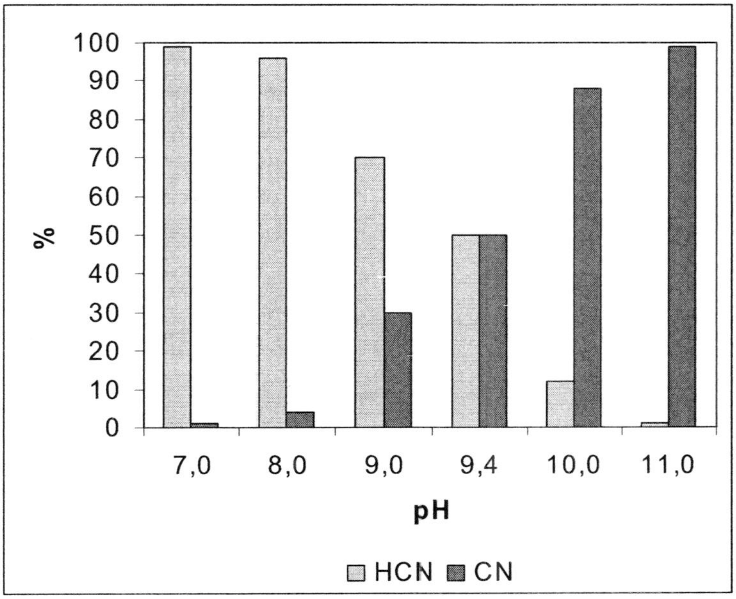

Fig. 1. CN^- and HCN content in solution as a function of pH of the medium (%).

This is the reason why chemical and microbiological methods ought to be combined. Such technologies exist so far only as patents (US Patent № 4,737.289, 1988. F.J.Castaldi, T.F.Trofe, G.C.Page, K.M. Adams; US Patent № 4,79,940,1988. F.J.Castaldi, T.F Trofe, G.C.Page, K.M. Adams). The essence of combined destruction methods comprises chemical transformation of a cyanide into a less toxic compound, such as, for example, a thiocyanate, followed by its microbiological destruction with an appropriate bacterium or activated sludge:

1st stage: $CN^- + S_x\text{-}S^{2-} = SCN^- + S_{x-1}\text{-}S^{2-}$
(pH 9.2-10), t = 70-80 ^{0}C, cationic surfactants (a catalyst).
Flotation with surfactants

2nd stage: $SCN^- + 2H_2O \rightarrow HS^- - NH_3 + CO_2$ *Thiobacillus,*
$NH_4^+ + 2O_2 \rightarrow NO_3^- + 2H^+ + H_2O$ *Nitrosomonas, Nitrobacter,*
pH 6.7 - 7.2, t = 25-35 ^{0}C
$HS^- + 2O_2 \rightarrow HSO_4$

The purpose of our study was to develop a combined technology for detoxification of cyanides and thiocyanates at high concentrations in industrial wastewaters and pulps.

2. Materials and Methods

2.1. Isolation of Bacterial Strains

Pure cultures of cyan-resistant strains capable of utilising cyanides and thiocyanates as a nitrogen source were selected by transferring enrichment cultures onto agar-containing synthetic media containing SCN^- as the sole nitrogen source. Out of the resultant 100 bacterial strains, two *Pseudomonas* strains (*Pseudomonas putida*, strain 21, and *Pseudomonas stutzeri*, strain 18) were selected on a liquid synthetic medium containing both SCN^- and CN^- as a nitrogen source. These two strains were found to be capable, both individually and as part of a consortium, of active cyanide and thiocyanate destruction. *P.putida* (ATCC 12633) and *P.stutzeri* (ATCC 17588) served as reference strains in order to compare physiological, biochemical and genetic characteristics of the strains selected.

2.2. Cultivation Media and Conditions

Cyan- and thiocyanate-resistant strains were isolated and cultivated on a medium containing the following mineral elements (g/l): $K_2HPO_4 \times 3H_2O$ – 3.0, $MgCl_2$ – 0.5, Na_2HCO_3 – 0.3, $FeCl_3$ – 0.01, $CaCl_2$ – 0.01. Also added were (g/l): NaSCN – 1.0, KCN – 0.01, potassium lactate – 4.0 or ethanol – 2%, agar-agar – 2 per cent, pH 9.0.

The destruction of cyanides and thiocyanates in industrial wastewaters and pulps was studied by cultivating the strains in 4-litre reaction vessels (filled with 2 litres of the growth medium); in the case of cyanide destruction, with mechanical mixing (150 rpm), at 25 °C and pH 9.2-9.4; in the case of thiocyanate destruction, with mechanical mixing (300 rpm) and aeration (2 l/min), at 30-34 °C and pH 8.8-9.0.

2.3. Analytical Methods

Bacterial growth was estimated in terms of cell number per 1 ml of the medium, using direct cell count under the microscope: a cell suspension (2 μl) was put onto a slide, covered with a 1x1-cm covering glass, and the edges were sealed with liquid paraffin. The cells were counted in 20 vision fields under a LUMAM I-1 microscope (magnification x 90 x 2.5 x 10). The number of cells in 1 ml of the suspension was calculated using the equation $X = 1.22 \cdot a \cdot 10^7$, where a is the mean value of cell number in 20 vision fields.

Cyanide-ion concentration was assessed in the pulp by titration with silver nitrate (Hamilton, 1932), and in the solution by using a crystalline ion-selective CRYTUR electrode, type 06-27, and an I-130 ionometer. Thiocyanate concentration was measured either by titration with $KMnO_4$ (5 N) (Hamilton, 1932) or analysed as ferrocyanate, using a colorimeter (Sörbo, 1957). Sulphite concentration was estimated by titration (Lurie, 1971). Trithionate, tetrathionate and thiosulphate content was determined by cyanolysis (Kelly, 1969). Sulphate concentration was measured using a Biotronik LC 500 liquid chromatograph. Sulphide was assessed as ZnS after precipitation (Trüper and Schlegel, 1964). NH_4 content was estimated by using the phenol-hypochloride colorimetric technique

(Weatherburn, 1967). Ethanol concentration was determined using an M-3700 gas-liquid chromatograph.

The structure of chromosomal DNA in the isolated and reference strains was studied by analyzing fragments yielded upon the restriction of native DNA with *Xba* I restriction endonuclease, using the technique of pulse electrophoresis (PS). The methods of native DNA isolation and restriction as well as the conditions of pulse electrophoresis were similar to those reported for the Gram-negative acidophilic chemolithoautotrophous bacterium *Acidithiobacillus ferrooxidans* containing, just as in the case of *Pseudomonas,* over 50 mol. % G + C in its DNA (Kondratyeva et al., 1995).

3. Results and Discussion

3.1. Screening of the New Strains

In this work, instead of using known strains and their adaptation to toxic compounds, we were looking for active strains resistant to such compounds in natural habitats. Our experiments followed the scheme:

1. Isolation of strains from samples of soils, sludge and industrial wastewaters.
2. Screening of the strains in terms of their resistance to CN^- и SCN^-.
3. Adaptation of the strains to increasing CN^- и SCN^- concentrations.
4. Using the strains for the purification of industrial wastewaters.

Microorganisms were isolated from samples of soils, silt and industrial wastewaters in the rewash discard of a gold-processing plant. 100 strains have been isolated, screened while growing on synthetic media for their resistance to CN^- (10 mg/l) and SCN^- (1 g/l) and, as a result, 40 bacterial strains resistant to 10 mg/l CN^- and 1 g/l SCN^- were selected. A consortium of two bacterial strains (*Pseudomonas putida*, strain 21, and *Pseudomonas stutzeri*, strain 18) resistant to 70 mg/l CN^- and 6 g/l SCN^- was obtained during the adaptation of the strains to increasing cyanide and thiocyanate concentrations. The strains we isolated were shown to differ from the type strains of the same species in physiological and genetic terms, namely, in the ability to use CN^- и SCN^- as a nitrogen source and in the structure of chromosomal DNA. The reference strains from the collection, *P.putida* B-2187 (IBFM) (ATCC 12633) and *P.stutzeri* B-975 (IBFM) (ATCC 17588) were incapable of CN^- and SCN^- destruction.

The study of mechanisms involved in the destruction of CN^- и SCN^- by the isolated strains has made it possible to establish that cyanide destruction yields NH_4^+, which is then used by the bacteria as a nitrogen source:

$$CN^- + 2H_2O \rightarrow NH_4^+ + CO_2$$

The mechanism of SCN^- destruction in *P.putida*, strain 21, differs from that in *P.stutzeri*, strain 18. Thiocyanate destruction by strain 18 yields thiosulphate, whereas, in the case of strain 21, thiosulphate is an intermediate product of thiocyanate destruction and polythionates are end products of SCN^- destruction:

$$SCN^- + H_2O \rightarrow HS^- + CNO^- \rightarrow NH_3 + CO_2$$
$$\downarrow$$
$$S_2O_3^{2-} + SO_3^{2-} \text{ (\textit{P.stutzeri}, strain 18)}$$
$$\downarrow$$
$$S_3O_5^{2-} + S_4O_6^{2-} \text{ (\textit{P.putida}, strain 21)}$$

The nitrogen of thiocyanate after transformation to ammonium catalysed by cyanase was also used by the bacteria of the consortium as a nitrogen source.

3.2. Adaptation of the Bacterial Consortium to CN^- И SCN^-

The isolated consortium of the strains was adapted to 70 mg/l CN^- (Fig. 2).

One can see that the rate of CN^- destruction was low (2.5-6.0 mg/l/day) during the first stages of adaptation to 10-20 mg/l of cyanide. It increased in the course of adaptation and became highest (35 mg/l/day) in the culture adapted to 70 mg/l CN. Further adaptation to even higher cyanide concentrations was pointless since, at pH 9.2, when bacterial growth was still possible, only 50 per cent of cyanide remained in the medium in a non-volatile state – CN^- (Fig. 1).

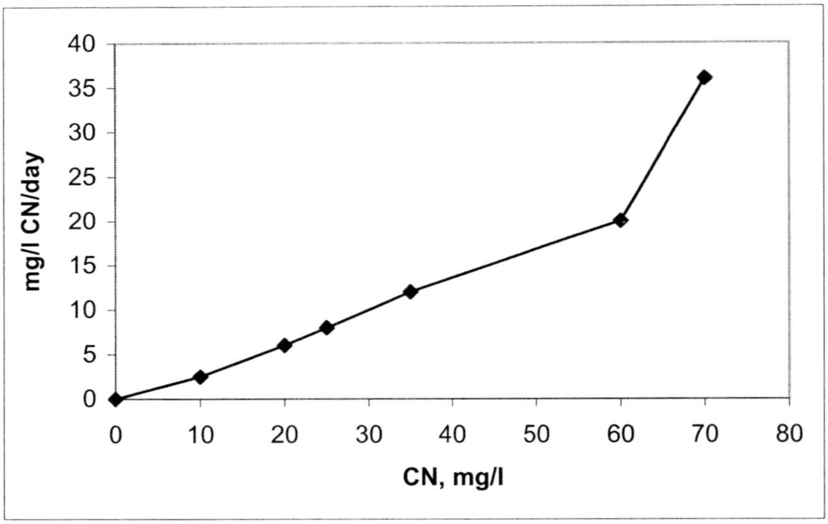

Fig. 2. The rate of CN^- destruction as a function of CN^- concentration to which the bacterial consortium has been adapted.

The bacterial consortium obtained as a result of adaptation was also capable of thiocyanate (4-6 g/l) destruction (Table 1).

Table 1. The rate of SCN^- destruction by the bacterial consortium of *P.putida* и *P.stutzeri* in different media

Medium	Original SCN^- content, g/l	Cell number, in 1 ml	Destruction rate, mg/l/hr
Liquid synthetic medium *	6.0 1.8 0.6	10^{11} 10^9 10^7	125 100 25
Pulp with a solid-to-liquid phase ratio of 1:5*	4.1 3.6	10^{10} 10^9	171 125
Pulp with a solid-to-liquid phase ratio of 1:2*	1.6 0.8	10^8 10^7	48 29

*- ethanol (4 g/g SCN) was added as a source of carbon and energy.

The rate of thiocyanate destruction depended on whether the bacterial consortium had sufficient quantities of a carbon source (an increase in the concentration of ethanol) and, as a result of this, on the number of bacterial cells in the medium.

Therefore a consortium of strains able to destroy high cyanide and thiocyanate concentrations was isolated.

3.3. Combined Bacterial-Chemical Technology

This consortium served as a basis for developing a combined bacterial-chemical technology for the detoxification of industrial wastewaters and pulps. Chemical pre-treatment was necessary to decrease the concentration of cyanide which could evaporate when the pH dropped and the medium was stirred in the course of microbiological destruction.

The combined bacterial-chemical method of CN^- and SCN^- destruction comprises two stages: (1) chemical and (2) microbiological, both safe with respect to environment. The content of cyanide and thiocyanates drops to the maximum permissible concentration of 0.05 mg/l in the course of destruction.

The **chemical stage** is required for cyanide destruction. Pulps or solutions are treated with an alkaline metal metabisulphite in the presence of copper sulphate as a catalyst, with stirring and aeration, in an alkaline medium (pH 10.7). The duration of treatment varies from 30 to 90 min depending on the initial concentration of the cyanide. In such conditions the residual cyanide concentration reaches 15 – 30 mg/l.

The **microbiological stage** is needed for the destruction of cyanides and thiocyanates by bacteria to reach the maximum permissible concentration. The process is conducted in the presence of phosphates and an organic carbon source needed for bacterial growth, in reactor with stirring and aeration. Cyanide is destroyed in conditions which prevent its evaporation: at pH 9.2 and 25-28 °C, with stirring. Thiocyanate is decomposed at a lower pH 8.5 – 8.8, with stirring and aeration, and the temperature being maintained above 34°C in order to prevent the bacteria eating by protozoa.

The method was tested in a laboratory reactor (Fig. 3) using industrial pulps (tailings from cyanidation sorption process of the bacterial oxidation products of gold-arsenic concentrate of the Olimpiadinskaya plant as a model). The wastewaters contained cyanides (200 mg/l), thiocyanates (4.5 g/l) and a number of heavy metals (iron, arsenic, etc.).

Fig. 3. Reactor for the bacterial destruction of CN^- и SCN^- in a dense pulp

The samples of cyanide wastewaters were detoxified by using sodium metabisulphite ($Na_2S_2O_5$) in a hermetic reactor with mechanical stirring and aeration. The rate of chemical detoxification was found to increase with a rise in the consumption of sodium metabisulphite (the ratio $Na_2S_2O_5/CN^-$) from 3 to 9 g/g CN^-. The residual content of CN^- dropped from 23.5 to 6.9 mg/l within 60 min when the consumption of sodium metabisulphite increased. Here the pH did not drop below 10.0, which prevented undesirable cyanide hydrolysis and made it possible to reduce CN^- losses. The addition of Cu^{2+} as a catalyst was shown to increase the rate of CN^- oxidation and to make it possible to reduce sodium metabisulphite consumption from 9 to 5 g/g CN^-. The pH value of the pulp after the chemical treatment was within 9.8 – 10.0.

An alternative method of pulp chemical pre-treatment with sodium thiosulphate was also tested in order to convert CN^- to SCN^-. A cyanide pulp originally containing 80 mg/l CN^- and 400 mg/l SCN^- was treated with sodium thiosulphate, 5 g or 7 g $Na_2S_2O_3$/g CN^- for 15, 30, 45, 60 and 75 min. The optimal concentration of CN^- (30 mg/l) for further bacterial destruction was obtained upon treatment with 7 g $Na_2S_2O_3$/g CN^- for 60 min, whereupon SCN^- concentration increased to 740 mg/l (Fig. 4).

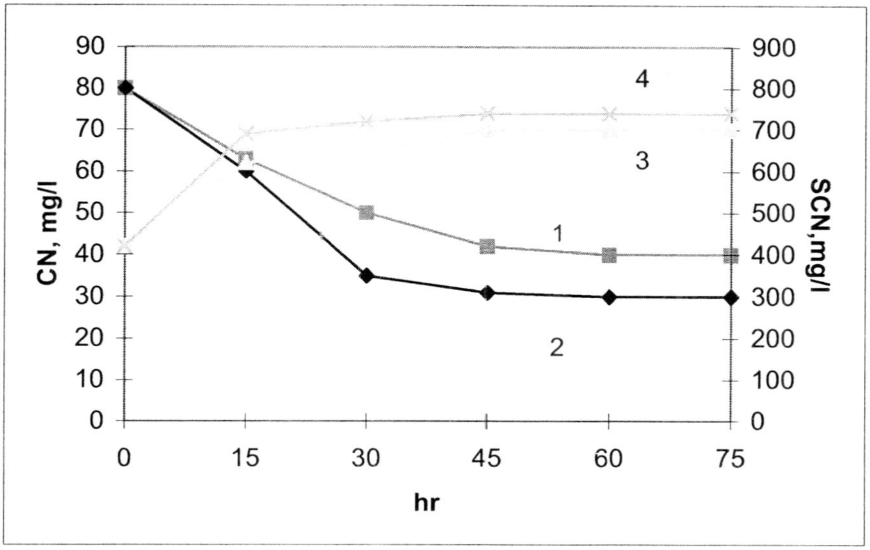

Fig 4. Effect of Na2S2O3 content and duration of the treatment on CN- and SCN- concentrations in the pulp: CN- mg/l (5 g Na2S2O3/g CN-);CN- mg/l (7 g Na2S2O3/g CN-);SCN- mg/l (5 g Na2S2O3/g CN-);SCN- mg/l (7 g Na2S2O3/g CN-)

The bacterial destruction of CN^- and SCN^- in dense pulps following the chemical treatment with sodium metabisulphite was effected by using a consortium of *Pseudomonas putida*, strain 21, and *Pseudomonas stutzeri*, strain 18. Lactate, sucrose or ethanol (the cheapest substrate) were utilized as a carbon and energy source, whereas CN^- and SCN^- served as a nitrogen source for the bacteria.

When the initial concentration of cyanide was 10 – 20 mg/l it was completely oxidised within 3 – 5 hrs. The rate of SCN^- destruction varied depending on its initial concentration. The maximal rate of SCN^- destruction (533 mg/l/hr) was reached with an initial concentration of 4.1 g/l and with the number of bacterial cells at 10^{10} in 1 ml (Fig. 5). Polythionates were the end products of thiocyanates destruction.

Thus the bacterial consortium under study was shown to be capable of CN^- and SCN^- destruction in dense pulps. It could also be effective for the purification of reversible solutions using immobilised cells. The processes of CN^- and SCN^- destruction can be intensified if the conditions are optimised.

CN^- and SCN^- destruction is based on their utilisation by bacteria as a nitrogen source. The process takes place only in the presence of carbon sources - in particular, ethanol - required for bacterial growth. Preliminary calculations indicate that the cost of the combined bacterial-chemical technology is lower than the cost of detoxification using the chlorination techniques employed by Mosquito Creek and Baker Mine plants (Smith and Mudder, 1991), but higher than in Giant Yellowknife where CN^- concentration was as low as 9 mg/l. Moreover, the bacterial-chemical technology makes it possible to destroy both CN^- and SCN^- at their high concentrations. Additionally, the technology of CN^- and SCN^- destruction by hypochlorite is ecologically harmful (Table 2).

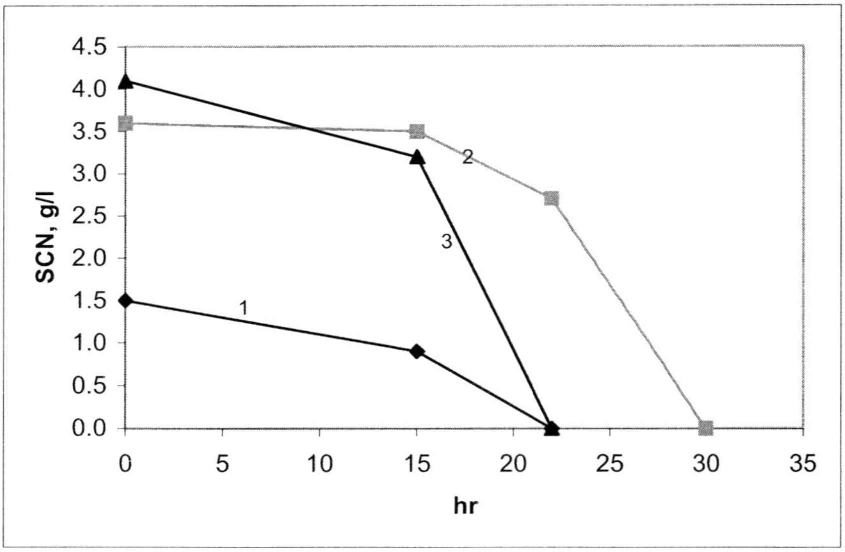

Fig. 5. Thiocyanate destruction by the bacterial consortium of *Pseudomonas putida*, strain 21, and *Pseudomonas stutzeri*, strain 18, in the pulp after bacterial leaching of gold at different SCN^- concentrations. The starting point of aeration is indicated with an arrow. The quantity of cells: 1 - 1.5 g/l SCN^- – 10^7 (cell/ml); 2 - 3.6 g/l SCN^- - 10^9 (cell/ml);3 - 4.1 g/l SCN^- - 10^{10} (cell/ml).

Table 2. Comparison of several parameters for the chemical and bacterial-chemical methods of detoxification of wastewaters containing cyanides and thiocyanates

Parameters	Industrial plants using alkaline chlorination			Combined bacterial-chemical technology
	Mosquito Creek	Baker Mine	Giant Yellowknife	
Concentration:				
CN^-, mg/l	310	2000	9	200
SCN^-, mg/l	-	-	-	4500
Cl_2 consumption, kg/m^3	5.45	13.0	0.12	0
Ethanol consumption, l/m^3	-	-	-	2
Cost / m^3 (\$)	3.94	9.07	0.106	2.5

4. Conclusions

1. The isolated consortium of *P. stutzeri*, strain 18, and *P. putida*, strain 21, was shown to be capable of destroying cyanides and thiocyanates at their high concentrations in various media. These bacteria utilize cyanides and thiocyanates as a nitrogen source. Ethanol, as the cheapest substrate, can be used as a carbon source to grow these bacteria.
2. Destruction of CN^- and SCN^- by the bacteria yields ammonium, which they use as a nitrogen source. Thiocyanate is also decomposed to yield thiosulphate (*P. stutzeri*, strain 18) or polythionates (*P. putida*, strain 21).

3. The consortium of *P. stutzeri* (strain18) and *P. putida* (strain 21) was found to be capable, in extreme conditions (pH, pulp, heavy metals), of developing high rates of CN^- and SCN^- destruction reaching 4 mg/l/hr and 533 mg/l/hr.
4. A combined bacterial-chemical method of CN^- and SCN^- destruction is proposed. It involves the chemical treatment of CN^- with an alkaline metal metabisulphite in the presence of a copper catalyst, as well as the microbiological destruction of SCN^- and residual cyanide with the consortium of *P. stutzeri*, strain 18, and *P. putida*, strain 21, in the presence of phosphates and organic matter.
5. Preliminary calculations demonstrate potential advantages of the combined bacterial-chemical technology: profound detoxification of industrial wastewaters from CN^- and SCN^- to the maximum permissible concentration, lower costs, etc.
6. Solving the problem of the purification of industrial wastewaters from cyanides and thiocyanates can make a significant contribution to the protection of the environment from these toxic compounds in various industries.

REFERENCES

[1] F. J. Castaldi, T. W. Trofe, G. C. Page A. K. M. Adams. *US Patent № 4,79,940* (1988).

[2] F. J. Castaldi, T. W. Trofe, G. C. Page A. K. M. Adams. *US Patent № 4,737,289* (1988).

[3] E. M. Hamil'ton (Ed.): *A Manual for Cyanidation of Gold and Silver Ores*. Tsvetmetizdat, Moscow, 1932. 41 p. (in Russian).

[4] D. P. Kelly, L. A. Chambers A. P. A. Trudinger: *Anal. Chemistry*, 41 (7), 898 (1969).

[5] T.F. Kondratyeva, L. N. Muntyan, G. I. Karavaiko: *Microbiology*, 141, 1157 (1995).

[6] Y. Y. Lurie (Ed.): *Universal Methods of Waters Analysis*. Hymia, Moscow, 1971, 375 p. (in Russian).

[7] A. Smith, T. Mudder: *Cyanide Chemistry and Treatment*. Mining. J. Books Ltd., London, 1991. 327 p.

[8] B. Sörbo. *Biochem. Biophys. Acta.*, 23, 412 (1957).

[9] H. G. Trüper A. H. G. Schlegel: *Ant.van Leev.*, 30, 225 (1964).

[10] J. L. Whitlock: *Geomicrobiology Journal*, 8, 241 (1990).

In: New Research on the Environment and Biotechnology ISBN 1-60021-042-2
Editors: A. Kuznetsov and G. Zaikov, pp. 135-146

Chapter 15

DISTINCTIVE FEATURES OF SULFATE-REDUCING BACTERIA FROM HYDROCARBON-BEARING RESERVOIRS HAVING AN ELEVATED TEMPERATURE OR SALINITY OF STRATAL WATERS

E. P. Rozanova*[1], *A. E. Ivanova and E. V. Belyakova
Winogradsky Institute of Microbiology of the Russian Academy of Sciences, Prospect 60-let Oktyabrya, 7/2, Moscow 117312, Russia.

ABSTRACT

This paper gives a detailed and generalized description of thermophilic sulfate-reducing bacteria (SRB), the representatives of various taxa, isolated by the authors from high temperature hydrocarbon-bearing reservoirs with different ecological conditions. Due to the demonstrated specific features, viz. the use of various electron donors and acceptors, fermentation of some organic compounds, and the temperature and salinity ranges of growth, these SRB are able to develop in flooded oil reservoirs with varying ecological conditions. Here we described two novel SRB isolates, binary hyperthermophilic association of archaea (98-A), and moderately halophilic strain 11, isolated from oil fields with high temperature and salinity of stratal waters. The former utilizes only lactate and pyruvate in the presence of sulfate or thiosulfate at temperature range from 52 to 87°C; one of its members is closely related to the *Archaeoglobus* genus, the other – to unculturable methanogenic archaea. The latter reduces sulfate in media with 4 - 23% NaCl (at the optimum of 8 – 10 %) under temperatures across the range 25°C – 50°C. Strain 11 able to grow autotrophically on H_2/CO_2. and oxidize a broad range of organic substrates. In the processes of organic compounds oxidation coupled with sulfate reduction the formation of acetate takes place, but butyrate is completely decomposes to CO_2.

[1] E-mail: epetroz@mail.ru; Fax: (095) 135 65 30.

Key words: hydrocarbon reservoir, sulfate-reducing bacteria, thermophile, halophile, complete oxidizer, ecological conditions.

The SRB are a physiological assemblage of anaerobic microorganisms with the unique ability of reducing sulfate to produce sulfide. SRB are mainly responsible for the terminal mineralization of organic material in anoxic sulfate-reach environments [1].

The SRB communities of deep subterranean ecosystems, hydrocarbon reservoirs in particular, have attracted a great attention in connection with the establishment of the lower boundary biosphere and the search for the new taxa. The growth of SRB in oil-bearing deposits has resulted in reservoir souring, the reduced permeability of the formation strata, corrosion of metallic facilities. Thus, the distribution and diversity of SRB in hydrocarbon fields is of theoretical and practical interest. The research results of such studies can be used to develop strategies for the prevention and control of H_2S production.

In practices of the exploitation, oil strata are flooded by river or seawater or a mixture of the surface and stratum waters, resulting in the appearance of ecotopes with different temperature, water salinity, and contents of sulfates and surface- or oil-originated organic compounds [2-4]. Thus, exploitation of the oil strata produces heterogeneous conditions for development of SRB.

For the whole period of investigation of different oil fields a lot of SRB were isolated; about 1/3 of them were thermophilic or halophilic bacteria [5]. The true halophilic species capable to complete oxidation of organic substrates were not known; only a few thermophilic representatives were carried out this process.

The SRB, completely oxidizers, from exploited terrigenic hydrocarbon deposits having an elevated temperature or/and salinity of stratal waters were in focus of the present investigation. This article emphasizes their metabolic flexibility and high adaptation potential as well as the importance of geochemical characteristics of hydrocarbon reservoirs in understanding their presence and activity in such extreme subterranean ecosystems.

The purpose of the present work is to summarize and discuss ecologically important properties of some recently described thermophilic species and characterize the new isolates from hydrocarbon reservoirs on the base of their phenotypic traits and the results of 16S rRNA gene sequencing.

MATERIALS AND METHODS

The metabolic and taxonomic diversity of SRB as well as physicochemical conditions of their habitats were studied during several expeditions to the hydrocarbon fields with high temperature or/and salinity of stratal waters. Recently thermophilic sulfate-reducing bacteria (*Desulfotomaculum kuznetsovii, Desulfotomaculum salinum, Desulfacinum subterraneum*) described by us and new SRB cultures (98-A and strain 11) were obtained from geographically distant high temperature oil and gas fields (Table 1) [2, 6-13].

Table 1. Thermophilic and halophilic sulfate-reducing bacteria from hydrocarbon-bearing reservoirs

Fields[1)], ecotopes, samples	Deposits, deph,m	Tempera-ture, °C	Minerali-zation, g/l, type water	SO_4^{2-}, mg/l	H_2S[2)], mg/l	Microorganisms[3)], strains	References
White Tiger oil field, Vietnam, anular space of injection well, water	Precenozoic, crystalline basement, >4000	30-55	35.5 (sea)	2620	0.02	Desulfacinum subterraneum (T), INMI 101[(T)]	2, 6
Producing well, water	Early Miocene, >4000	106	25.5 (sea and stratal)	932	2[4)] sm^3/m^3	Archaeoglobus sp. (T) (in binary association), INMI 98 A	Results of this study
Producing well, oil and water						Alkane-oxidizing consortium (T), INMI 98 P	2
Ust'- Balykskoye oil field (West Siberia, Obskaya Group). Producing well, water	Cretaceous, 2100	63-70	16.1 (stratal)	0	0	Desulfutomaculum salinum (T), RKM B-1379	7, 8
Igrimskoye gas field (West Siberia). Producing well, water	Middle Jurassic, 1620	59	9.8 (stratal)	0	0	Desulfutomaculum salinum (T), RKM B-1492[(T)]	7, 8
Marine structure of Peschaniy island oil field, Apsheron peninsula. Producing well, water	Tertiary, Productive suit, 3291	84	15.1 (stratal)	362	114.50	Thermodesul-fobacterium thermo-philum (T) INMI 7[(T)]	9, 10
Colhida lowland bitumen deposits (Georgia). Mineral spring	Cretaceous, 3250	63	2.8 (stratal)	285	1.34	Desulfotomaculum kuznetsovii (T) RKM B-1805[(T)]	11, 12
Uzen' oil field, near-bottom zone of injection well	Jurassic 1300	55-65	16.8	2850	75	Desulfotomaculum kuznetsovii, INMI 2175	13
Persian Gulf area, oil field. Producing well, water	Tertiary, <2600	<90	213.0 (stratal and small quantity of injected)	33	0	Desulfovermiculus halophilus (H) INMI 11[(T)]	Results of this study

1) Water-flooded oil fields, terrigenous rocks;

2) Intensity of sulfate-reduction in the samples; mg S. l-1 . day-1: N 1 – 3,03; N 6 –83,5; N 7 – 3,07; (data reproduced from [2, 12])

3) T – thermophilic, H – halophilic, (T) – type strain

4) In the stratal gas

Samples from two different oil fields were used for enrichment and isolation of new thermophilic and halophilic sulfate-reducing bacteria during this study. Strain 11 ("*Desulfovermiculus halophilus*") was isolated from a thermophilic anaerobic enrichment culture growing at 37°C. The initial inoculum was a sample of production water from oil field located in Persian Gulf area. The enrichment cultures were recovered from the highest positive dilutions of sulfate-supplemented medium. Pure cultures of sulfate-reducing bacteria were obtained from liquid enrichments by serial dilution in medium of the same composition, solidified with 1.5% (w/v) agar (Difco), and consequent isolation of single colonies.

Binary hyperthermophilic association 98-A (archaea) was obtained from a sample of formation water of high temperature White Tiger oil field, Vietnam. Enrichments were grown at 65°C. Culture did not produce colonies on any of solidified media therefore for purification procedure we used a direct dilution method in liquid medium supplemented with antibiotics. Purity was checked by light microscopy and by verification the absence of growth in a medium containing yeast extract (0.5 g/l) and glucose (1 g/l).

A bicarbonate buffered anaerobic saltwater medium [14] with some modifications was used for cultivation of both cultures. For cultivation of halophilic strain 11 increased concentrations of NaCl and $CaCl_2$ x $2H_2O$ (100 and 1 g/l respectively) were used; for binary association 98-A 0.5 g/l $CaCl_2$ x $2H_2O$ was added to the medium. Utilization of substrates by SRB was determined by monitoring growth, substrate depletion and product formation. For growth on alternative electron acceptors 10 mM , thiosulfate, 2 mM sulfite, 5 mM nitrate, or 1% elemental sulfur replaced the sulfate.

The temperature optimum was determined in appropriate medium for each culture, and a temperature range of 15-92°C. In order to determine the NaCl, $MgCl_2$, or $CaCl_2$ range for growth, the concentration of one of these salts was varied while the concentrations of the other inorganic compounds were maintained.

Morphology of new isolates was observed with a phase-contrast microscope. Isolayion and purification of genomic DNA, G+C content of the DNA determination and amplification and sequencing of 16S rDNA was carried out as described previously [6]. The cell density was determined by direct counting in a light microscope. Hydrogen sulfide production was determined by colorimetric method [15]. Concentration of molecular hydrogen, methane, lower fatty acids and alcohols was analyzed by means of GC and GLC as before [7]. Phylogenetic analysis was carried out by Tourova N.P.

Results and Discussion

Taken together, our studies demonstrated the occurrence of thermophilic and halophilic SRB in high-temperature hydrocarbon-bearing reservoirs – oil, gas, and bitumen deposits (Table 1) with substantially varying temperature, stratal waters salinity, and the sulfate and sulfide content. Oil fields were flooded with injected seawater at the White Tiger oil field, or with mixed sea and stratum waters at the Peschaniy island offshore and the Persian Gulf oil field, or fresh water in the Westeren Siberia fields at the early exploitation stages. Flooding of these oil fields resulted in the intensification of water exchange in the strata and the appearance of zones with different temperature and water mineralization. In the near-bottom zones of White Tiger injection wells temperature decreased to 50°C, and in the Western

Siberia oil field - below 20°C. Waterflooding of White Tiger was followed by increasing of stratal water mineralization up to 35.5 g/l (starting values – 5-7 g/l) and in Western Siberia oil fields lowering to several hundreds mg/l (starting values – 15 – 20 g/l). The stratal water was enriched by sulfates that entered with seawater. As to the Western Siberia oil fields, sulfide-reached rocks were a source of sulfates in the stratum water. Sulfides were oxidized by oxygen of air dissolved in injected water bulks. Thus, flooding facilitated the processes of sulfate reduction in the high temperature oil fields (Table 2), which occurred from near-bottom zones of injection wells deep into oil strata in the West Siberian and White Tiger oil fields (table 2) [2-4]. Hydrogen sulfide accumulated in stratal waters only after exhausting of Fe^{2+} dissolved. The presence of SRB in the stratal waters (Table 1) was the first indication of sulfate reduction.

Table2. Number of SRB and rate of sulfate reduction process in the water-flooded high temperature oil fields

Oil field	Habitat	Temperature, ^{0}C	Mineralization, g/l	Fe^{2+}, mg/l	SO_4^{2-}, mg/l	H_2S (in water), mg/l	Number of SRB$^{1)}$, cells/ml	Rate of sulfate reduction processes$^{2)}$, μg S^{2-} l^{-1} day^{-1}	References
Western Siberia oil fields of Obskaya group: Mykhpayskoye, cretaceous	Producing wells (6 samples)	> 60	13.6-17.9	Up to 50	1.5-6.7	0	0-10^1	0.03-9.50	3
Mamontovskoye, cretaceous	Producing wells (5 samples)	70-73	9.6-21.0	Up to 50	9.3-17.2	0-0.1	10^0-10^1	0.10-9.50	4
White Tiger, Vietnam Basement	Near-bottom zones of injection wells (2 samples)	53-60	35.5	< 50	2567-2587		10^1	12.94	2

1) In the media with Na-lactate
2) At 70^0 C

The data on the rate of sulfate reduction pointed to the active development of SRB, the representative of which was *Desufotomaculum salinum*, in flooded strata of the West Siberian oil field (Cretaceous deposits). However, the origin of these microorganisms has been unclear as yet. The presence of the same spore-forming SRB species in the water-flooded Ust'-Balykskoe oil field and non-flooded Igrimskoe gas field led us to suppose that these bacteria are indigenous in such habitats. Possibly, they played the role in disappearance of sulfates in the stratal waters during their formation and precipitation of sulfides in stratal rocks. Undoubtedly, thermophilic SRB participate in the formation of biogenic hydrogen sulfide at the Uzen' field (the Peschaniy island offshore) and the bitumen deposits in Colhida, as evidenced from the measured sulfate reduction rates (Table 1). The possibility of biogenic sulfate reduction development in the Early Miocene deposits of the White Tiger oil field, where the thermophilic *Archaeoglobus* sp. and alkane-oxidizing consortium of SRB are present, has been supported by the significant rates of sulfate-reduction in near-bottom zones of injection wells (Table 2).

Among the isolated thermophilic SRB, we found the representatives of phylogenetically diverse taxa, such as the gram-positive spore-forming bacteria *Desulfotomaculum kuznetsovii* and *D. salinum*; the gram-negative bacteria *Thermodesulfobacterium thermophilum*, *Desulfacinum subterraneum,* and the archaeon *Archaeoglobus* sp. (Table 1).

The studied thermophilic SRB oxidized various organic compounds (Table 3) [6-11]. The majority of microorganisms that are listed in Table 3 completely decomposed organic substrates to CO_2 with exception for *T. thermophilis*, the incomplete oxidizer. Some of oxidized substrates were originated from oil, such as the lowest alcohols and fatty acids. Also, these substances might appear due to aerobic bacterial hydrocarbon oxidation in near-bottom zones of injection wells. Fatty acids with chain length C_2-C_{16} were obtained by passing mineral medium through a porous reservoir setup containing hexadecane and incubated with aerobic hydrocarbon oxidizing bacteria [16]. Amino acids, the substrates for *D. subterraneum* (Table 3), are components of biomass formed in oil strata due to bacterial activity and are released in the result of cell lysis. Similarly, some other organic acids (lactate, pyruvate etc.) have a biological nature and can be utilized by SRB. Finally, the petroleum components, n-alkanes with the chain length C_{12}-C_{20}, may serve as substrates for oil-stratum SRB developing at high temperature. We isolated from the Miocene deposits of the White Tiger oil field the association of SRB capable to decompose these alkanes at 70-80°C (Table 1). These bacteria also oxidized butyrate and capronate during sulfate reduction. Alkane-oxidizing SRB were previously found in marine hydrocarbon-contaminated habitats [17].

Four of the investigated thermophilic SRB strains were able to grow lithoautotrophically using H_2/CO_2 and one strain could grow under lithoheterotrophic conditions in the presence of H_2/CO_2 and acetate or yeast extract. Hydrogen was detected as a component of stratum gas from the Western Siberia oil field, a natural habitat of the bacterium *D. salinum,* which oxidizes H_2 in the process of sulfate reduction. The accumulation of H_2 as well as some organic substrates (Table 3) may result from fermentation of bacteria biomass in the strata. Besides this, various organic substances enter with superficial water into flooded oil strata. In the absence of sulfates, the studied thermophilic SRB fermented the certain of organic substances, such as pyruvate, fumarate and yeast extract components (Table 3).

The sporulating SRB, having the specialized dormant and stress-resistant forms, survive unfavorable conditions for prolonged periods being in a resting state. It is possible that *D. salinum* exists in the form of spores in H_2S and SO_4^{2-} - free strata of the Igrimskoye gas deposit. Besides sulfates, thermophilic SRB were capable to utilize as electron acceptors the other reduced sulfur compounds, such as thiosulfate, sulfite, or elemental sulfur and Fe III (Table 3). Such compounds may appear in flooded oil reservoirs in the result of sulfide oxidation or Fe III by oxygen of air dissolved in injected water.

Table 3. Properties of sulfate-reducing bacteria, isolated from hydrocarbon-bearing deposits.

Characteristics	Thermophilic SRB	Halophilic SRB				
	D. subterraneum INMI 101^(T) [6]	D. kuznetsovii RKM B-1805 [11]	D. salinum 1) RKM B-1492 [7, 8] 2) RKM B-1379 [7, 8]	T. thermophilum INMI 7 [9, 10]	Binary association including Archaeoglobus spec. INMI 98 A [Results of this study]	*Desulfovermiculus halophilus* INMI 11 [Results of this study]
Morphology	Non-motile oval	Sporeforming motile rods	Sporeforming motile rods	Rods	Irregular coccoid	Short rods or long irregular curved cells
G+C, mol %	60.8	49.0	1) 50.8 2) 50.6	38.0	N.D.*	55.2
Optimal salinity (range salinity), g/l	5(0-50)	0-1 (0-3)	1)1 (0-6) 2) 0.5 (0-4.5)	1 (0.5-20)	20-30	110 (40-230)
Optimal temperature, ^{0}C	60 (45-70)	60-65 (50-85)	1) 60-65 (45-75) 2) 60-65 (45-70)	65 (45-85)	65 (55-92)	37 (25-50)
Optimal pH	7.2-7.4 (6.5-8.4)	7.0-7.2	1) 7.0 (6.0-8.5) 2) 7.0 (6.5-8.5)	N.D.	6.8-7.2	7.2 (6.0-8.5)
Oxidation of organic compounds	Complete	Complete	1), 2) Complete	Incomplete	Complete	Complete (butirate)
Organic electron donors (with SO_4^{2-})	Lactate, pyruvate, fumarate, malate, ethanol; fatty acids (C-number): C_1-C_6,C_{16}; alanine, serine, cysteine, yeast extract	Lactate, pyruvate, fumarate, malate, succinate, methanol, ethanol, propanol, butanol; fatty acids (C-number): C_1 –C_7, C_{17}, C_{18}	1), 2) Lactate, pyruvate, fumarate, malate, succinate, methanol, ethanol, propanol, butanol; fatty acids (C-number): C_1 –C_6, C_{16} ; yeast extract	Lactate, pyruvate, formate (last in the presence of acetate)	Lactate, pyruvate	Lactate, pyruvate, fumarate, malate, succinate, ethanol, formate, propionate, butirate, alanine

Table 3. Properties of sulfate-reducing bacteria, isolated from hydrocarbon-bearing deposits. (Continued)

Characteristics	Thermophilic SRB	Halophilic SRB				
	D. subterraneum INMI 101$^{(T)}$ [6]	D. kuznetsovii RKM B-1805 [11]	D. salinum 1) RKM B-1492 [7, 8] 2) RKM B-1379 [7, 8]	T. thermophilum INMI 7 [9, 10]	Binary association including Archaeoglobus spec. INMI 98 A [Results of this study]	*Desulfovermiculus halophilus* INMI 11 [Results of this study]
Morphology	Non-motile oval	Sporeforming motile rods	Sporeforming motile rods	Rods	Irregular coccoid	Short rods or long irregular curved cells
Growth on H_2/CO_2 lithoautotrophic lithoheterotrophic	 + ** +	 + +	 1), 2) + 1), 2) +	 - +	 - -	 +
Fermentation	Pyruvate, yeast extract	Pyruvate, fumarate	1), 2) Pyruvate, (fumarate-slowly)	-	-	Lactate, pyruvate, fumarate
Other electron acceptors	SO_3^{2-}, $S_2O_3^{2-}$, S^O	SO_3^{2-}, $S_2O_3^{2-}$ ***, Fe III ****	SO_3^{2-}, $S_2O_3^{2-}$,x S^O, Fe III ****	SO_3^{2-}, $S^2O_3^{2-}$	SO_3^{2-}	SO_3^{2-}, $S_2O_3^{2-}$, S^O, NO^{3-}

* Data absent; ** + - Positive results; - Negative results; *** Reduction and dismutation;

**** Results of this investigation: *D. salinum* – reduction Fe III, but cannot grow; *D. kuznetsovii* reduction Fe III and grow.

As follows from the comparison of natural habitats of thermophilic SRB (Table 1) and their physiological properties (Table 3), these bacteria, as a rule, are well adapted to environmental conditions in hydrocarbon-bearing deposits. The demonstrated tolerance to widely varying temperature, salinity, and pH provides for the existence of SRB under heterogeneous conditions in flooded oil reservoirs.

As yet, the functions of a binary association of archaea including *Archaeoglobus* sp. have not clear at all. It is possible that this association can exist at temperatures > 92°C under the elevated pressure typical for the strata of the White Tiger oil field [2]. Little has been known about the role of the second member of the association related to unculturable methanogenic archaea [18].

Hence, due to a variety of metabolic potentialities, the studied thermophilic SRB could survive under altered conditions in water-flooded oil reservoirs.

Below we consider properties of moderately halophilic SRB designated as strain 11. Halophilic SRB strain 11 is a gram-negative bacterium with rod-shaped or curved, sometimes vibriod, cells of 0.6-0.8 µm × 1.5-12 µm in size. The bacterium grows in media with the NaCl content from 40 g/l to 230 g/l with the optimum at 80 g/l (Fig. 1). The optimal concentrations of $CaCl_2$ x $2H_2O$ and $MgCl_2$ x $6H_2O$ in growth media are 1 g/l and 3 g/l, respectively (4.57 mM Ca^{2+}- Fig. 2). This bacterium grows at the temperature range 25-50°C and the optimum at 37°C. A distinct feature of this halophilic bacterium is the ability to reduce sulfate and grow lithoautotrophically by metabolizing H_2/CO_2 (Table 3). Another distinct feature of the bacterium is the ability to decompose butyrate during sulfate reduction without accumulation of acetate. This feature is typical for the group of complete oxidizers of organic compounds, which includes the majority of thermophilic SRB studied by us. Accumulation of acetate occurs when the bacterium grew in media with other organic substances, e.g., propionate, malate, fumarate, succinate, alanine The growth of the strain 11 on a medium with butyrate under optimal temperature, pH, and salinity was slower than on a medium with propionate.

As already reported [1,7], SRB that are able to completely oxidize some organic compounds to CO_2 decompose other organic substrates with the accumulation of acetate. This fact is thought to be associated with differences in the growth rates and acetyl-CoA yields during decomposition of various substrates. With this in mind, it is possible to explain the peculiarities that we observed in oxidation of butyrate and the other organic substrates by the halophilic SRB strain 11.

Hence, we affiliate the studied SRB strain 11 to the complete oxidizers. The ability of this bacterium to grow lithoautotrophically is in consistence with this affiliation and points to the presence of the Wood pathway enzymes responsible for acetyl-CoA decomposition to CO_2. One of these enzymes is CO-dehydrogenase with a reversible type of action [19]. The revealed formation of methane at little amount (286 nm CH_4 per 100 ml of the medium) during the lithoautotrophic growth on H_2/CO_2 is an additional support to regard the studied strain as a representative of the complete oxidizers [19].

The C+G content in DNA of the strain 11 cells was 55.2 mol %. On the basis of 16S rRNA sequence data, strain 11 falls into the delta subdivision of the *Proteobacteria*, forms a separate branch and clusters with members of the family *Desulfohalobiaceae* (as proposed by Kuever et al.) [20]. This strain is sufficiently different from other members of the known taxa of this family to be classified as a representative of a new genus; the name *Desulfovermiculus halophilus* gen. nov., sp. nov. is proposed.

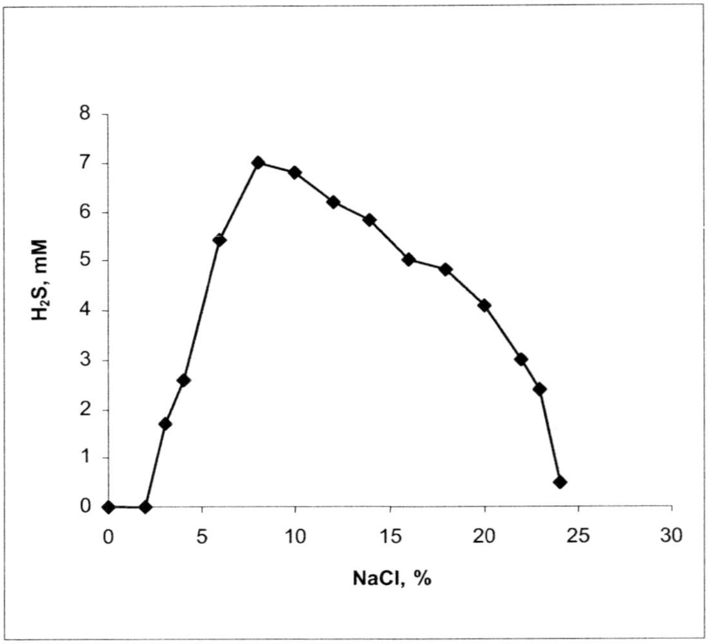

Fig. 1. Effect of NaCl concentration on production of sulfide *Desulfovermiculus halophilus* strain 11 grown on malate and sulfate at 37 ^{O}C.

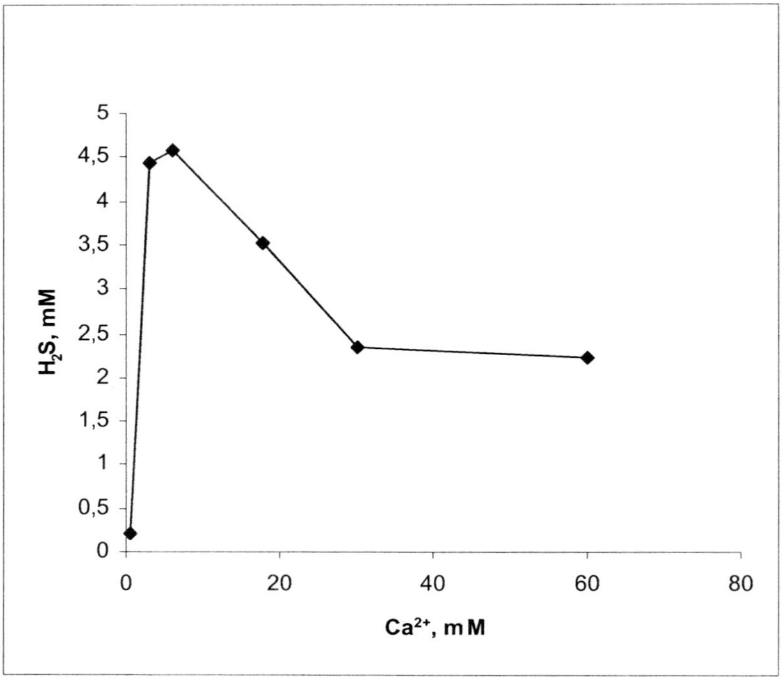

Fig. 2. Effect of Ca^{2+} concentration on production of sulfide *Desulfovermiculus halophilus* strain 11 grown on malate and sulfate at 37 ^{O}C

Regarding the possibility of the studied halophilic SRB strain 11 functioning in high-temperature oil-bearing strata at the oil field located in Persian Gulf area (Table 1), it is worth to note that salinity of stratum water is in the range, which permits the growth of this organism. However, the upper temperature limit for the growth of the bacterium studied does not exceed 50°C. Unfortunately, we have no any data on the temperature of strata at time of sampling; it is possible that water injection from the surface resulted in cooling of habitats where this bacterium occurred. It is possible to suggest that the real habitat for the microorganism studied are stratum water reservoirs located on the land, e.g., oil settling tanks or system of injecting water preparation. It is probable that this bacterium enters deep oil strata with injected water.

Increased salinity of stratal brines enriched with calcium chloride, as well as moderate temperatures, are peculiar to oil fields of the Volga-Ural region (Russia). The environmental conditions in these oil reservoirs are suitable for microorganisms like the studied halophilic SRB, which tolerates high concentrations of sodium chloride and calcium chloride and low pH, typical of stratal brines. As already reported in the literature, flooding of terrigenous reservoirs located in the Volga-Ural region leads to the appearance and gradual increasing of biogenic hydrogen sulfide in stratal brines. With the use of labeled $^{35}SO_4^{2-}$, it was possible to detect a slow development of biogenic sulfate reduction amounted tenths μg $S^{2-} l^{-1} \cdot day^{-1}$ in stratal water with the salinity of more than 200 g/l [21].

Stratal waters in numerous deep subsurface oil reservoirs are highly saline and hot. However, the SRB that are able to develop in such conditions have not so far been found, although theoretically they may exist.

Recently, the origin of microorganisms occurring in subsurface ecosystems has been a subject for discussion. Some researchers believe that hyperthermophiles belong to a common deep subterranean biosphere [22, 23]. We agree with this viewpoint and think that thermophilic microorganisms, having the similar properties, are widespread in various ecosystems with the same conditions. The examples are given in Table 1. On the other hand, microorganisms are distributed in the exploited oil fields due to flooding with seawater. Thermophilic SRB *Desulfotomaculum kuznetsovii* were found in cold seawater and flooded high-temperature oil reservoirs [24]; the existence of thermophiles in oceans is associated with hot water streams flowing from breaks or cracks in the Earth crust.

Yet, the upper temperature boundary for life has not been established. Fe III – reducing microorganism (archaeon) incubated for up to 2 hours at 130°C still grew when trasferred to fresh medium 103°C [25]. To the present time, representatives of the sulfate-reducing bacteria, which can grow and develop at temperatures above 100°C, have not been found, as well as extremely halophilic SRB. Further search for thermophilic and/or halophilic SRB is promising.

REFERENCES

[1] F. Widdel: in A.J.B. Zehnder (Ed.): *Biology of Anaerobic Microorganisms*. John Wiley a. Sons, New York etc.,1987. 469-584.

[2] E.P. Rozanova, I.A. Borzenkov, A.L. Tarasov, et al.: *Microbiology* (Engl. Tr.), 70 (1), 118 (2001).

[3] S.S. Belyaev, E.P. Rozanova, I.A. Borzenkov, et al.: *Microbiology* (Engl. Tr.), 59 (6), 1075 (1990).

[4] E.P. Rozanova, A.S. Savvichev, YuM. Miller, M.V. Ivanov: *Microbiology* (Engl. Tr.), 66 (6), 852 (1997).

[5] T. N. Nazina, S.S. Belyaev: in V.F. Galchenko (Ed.): *Proceedings of Winogradsky Institute of Microbiology*. Vol. 12. Nauka, Moscow, 2004. 289-316 (in Russian).

[6] E.P. Rozanova, T.P. Tourova, T.V. Kolganova et al.: *Microbiology* (Engl. Tr.), 79 (4), 536 (2001).

[7] E.V. Belyakova, E.P. Rozanova: *Microbiology* (Engl. Tr.), 73 (2), 284 (2004).

[8] T. N. Nazina, E.P. Rozanova, E.V. Belyakova, A.M. Lysenko et al.: *Microbiology* (submitted), 74 (4), (2005).

[9] E.P. Rozanova, A.L. Khudiakova: *Microbiology* (Engl. Tr.), 43 (6), 1069 (1974).

[10] H.G. Trüper: *Int. J. Syst. Evol. Microbiol.*, 53, 927 (2003).

[11] T. N. Nazina, A.E. Ivanova, L.P. Kancaveli, E.P. Rozanova: *Microbiology* (Engl. Tr.), 57 (5), 823 (1988).

[12] L.P. Kancaveli: Subcerface Water of Colhida Lowland in Connection with Genesis of Hydrogen Sulfide of Microbiological Origin. *Ph. D. Thesis*. Djavahishvili University, Georgia, Tbilisi, 1992. 24 p. (in Russian).

[13] A.E. Ivanova: Hydrocarbon-Oxidizing and Sulfate-Redusing Bacteria From High Temperature Oil Stratum. *Ph. D. Thesis*. Institute of Microbiology RAS, Moscow, 1998. 24 p. (in Russian).

[14] K F. Widdel, T. Bak: in A. Ballous et al. (Eds.): *The Prokaryotes* (2-nd ed.), Vol. 4. Springer-Verlag, New York etc., 1992. 3352-3378.

[15] H.G. Trüper, H.G. Schlegel: *Antonie van Leeuwenhoek*, 30 (3), 225 (1964).

[16] E.S Kulik, Yu. P. Somov, E.P.Rozanova: *Microbiology* (Engl. Tr.), 54 (3), 304 (1985).

[17] J. Heider, A.M. Spormann, H.R. Beller, F. Widdel: *FEMS Microbiology Reviews*, 22, 459 (1999).

[18] T.V. Kolganova, B.B. Kuznetsov, T.P. Tourova: *Microbiology* (Engl. Tr.), 71 (2), 243 (2002).

[19] R. Schauder, B. Eikmanns, R.K. Thauer et al.: *Arch. Microbiol.*, 145 (2), 162 (1986).

[20] J. Kuever, F.A. Rainey, F. Widdel: "Family Desulfohalobiaceae Fam. Nov.: in G.M. Garrity. (Ed.): *Bergey's Manual of Systematic Bacteriology*, 2nd edn. Vol. 2. Springer-Verlag, New York (in press).

[21] I.A. Borzenkov, A.L. Tarasov, E.I. Milekhina, S.S. Belyaev: in G. E. Zaikov (Ed.): *Biotechnology and the Environment Including Biogeotechnology*. Nova Science Publishers Inc., New York, 2004. 171-178.

[22] K.O. Stetter, R. Huber, E. Blöchl et al.: *Nature*, London, 365, 743 (1993).

[23] R.K. Nilsen, J. Beeder, T. Thorstenson, T. Torsvik: *Appl. Env. Microbiol.*, 62 (5), 1793 (1996).

[24] M.F. Isaksen, F. Bak, B.B. Jørgensen: *FEMS Microbiology Ecology*, 14 (1), 1 (1994).

[25] Kashefi, D.R. Lovley: *Science* 301, 934 (2003).

In: New Research on the Environment and Biotechnology ISBN 1-60021-042-2
Editors: A. Kuznetsov and G. Zaikov, pp. 147-157

Chapter 16

IDENTIFICATION AND SOME ENZYMATIC PROPERTIES OF NEW THERMOPHILIC BACILLI

H. H. Panosyan[1,*], P. V. Tozalakyan[2], R. Paskauskas[3], A. Zvikas[3] and Y. G. Popov[1]

[1]Yerevan State University, 375025, Yerevan, Alex Manoogian st., 1, Armenia
[2]GEORISK Scientific Research Company, 375019, Yerevan, Baghramyan Ave., 24a, Armenia
[3]Institute of Botany, LT-08406, Vilnius, Zaliuju ezeru st., 49, Lithuania

ABSTRACT

Two thermophilic cultures of bacilli isolated from the soil samples at the site of geochemical and thermal anomalies in Dilijan city (Armenia) were identified down to species. The morphological, physiological and biochemical properties, as well as determination of 16S rDNA sequences of thermophilic strains were described. Data on some enzymatic activities (protease, CGT-ase, amylase, lipase, β-galactosidase) of the isolated strains were also reported.

Key words: thermophiles, geochemical and thermal anomalies, enzymatic activities, *Bacillus*

INTRODUCTION

Isolation and study of thermophilic bacteria from the sites of thermal anomalies is of great interest due to their potential for biotechnology [1, 2].

Microorganisms isolated from habitats, where high temperatures are combined with other factors such as alkaline or acidic pH, pollution by various organic and inorganic sources of carbon, nitrogen and other nutritional and/or toxic components are particularly promising.

---

* E-mail: physiol@ysu.am,Fax: (3741) 554-641

One can expect that they will have unusual characteristics and might serve as a valuable resource for application in novel biotechnological processes [2].

The events recorded in 1999 in the town of Dilijan (Armenia) can be described by the following sequence:

Landslip activation ⇨Damage of the sewerage system ⇨Formation of conditions favorable for growth of microorganisms ⇨Increase of soil temperature up to 70°C ⇨Development of thermophiles and other extremophiles at high pH level and in the presence of sewage.

Various extremophilic bacilli were isolated from this place [3]. Their potential for biotechnology should be related to their enzymatic properties.

Amylases are the principal components of mixtures used in the removal of starchy sizing agents from textiles and in the conversion of cereal grains to fermentable sugars.

Lipases are used in detergents for the improvement of the cleaning action as well as in food industry for flavor improvement and splitting fats without damaging sensitive constituents.

Cyclodextrins (cyclomaltooligosacharides, cyclic (1-4)-a-D-glucans, CDs) are produced from starch, amylose, amylopectin and other long-chain maltooligosaccharides by the action of *cyclodextrin glucantransferase* and represent a class of very important biotechnological products. They have been widely utilized in food, pharmaceutical, chemical, cosmetic and agricultural industries due to ability to form inclusion complexes with wide variety of chemicals and to alter the physical and chemical properties of these compounds [4].

β-galactosidase is of special interest for the treatment of milk to meet the needs of large percentage of the world population afflicted with lactose intolerance. Industrial applications necessitate the enzyme's immobilization, on which considerable investigation has been reported [5-11].

In this paper isolation of two thermophilic strains of bacilli from the site of thermal and geo-chemical anomalies (Dilijan, Armenia) and their genotypic and phenotypic characterization are described. Protease, CGT-ase, amylase, lipase, β-galactosidase activities of the isolated strains were studied, and their potential for biotechnology was estimated.

Materials and Methods

Study Site

Soil samples were taken from a depth of 1m at the site of thermal and geo-chemical anomaly recorded in Dilijan. The physical and chemical parameters of the soil were the following: pH - 8.4; Eh – 177 mV; conductivity of water extract – 423 mS/cm; mineralization of water extract - 423 mg/l. The temperature of soil at this site reached up to 70°C before sampling. These parameters were measured in field conditions by WTW-320 pH-meter and WTW-LF-95 conductometer.

Enrichment and Isolation

Soil suspensions were exposed to the thermal treatment at 80^0 C for 10 min to kill vegetative cells and to induce spore germination. 1 ml of the treated suspension was inoculated into nutrient broth and incubated with shaking at 56°C temperature. In case of turbid growth, cultural broth was spread on a fresh 2% (w/v) agar medium before incubation. Colonies of different morphology were identified and repeatedly streaked for purity.

Enrichment medium contained the following components (grams per liter in distilled water): NaCl - 5; $CaCO_3$ - 5; peptone - 5; glucose - 10; corn extract – 5; pH – 7,2-7,4. Solid medium was prepared by the addition of 2% agar [12].

Morphological Properties and Staining

Cellular morphology and appearance of motility were determined by phase contrast microscopy (PZO, Warszawa SK14). Gram stains were made by the common method [13]. The cells were stained for presence of endospores using Peshkov's method [13]. Endospores were photographed by Canon PC1049 digital camera combined with the Olympus light microscope model CH2 (1500×).

Physiological and Biochemical Properties

The temperature range for growth was determined after incubation of isolates at 5 to 80°C by 5°C intervals. The pH dependence of growth was tested at pH range of 5 to 12. The range of NaCl concentrations for growth was determined by adding 0 to 15% NaCl to the incubation medium.

Catalase activity was determined by bubble formation in a 3% hydrogen peroxide solution.

Anaerobic growth of the isolates was tested by growth in solid anaerobic stab cultures covered by paraffin layer.

Reduction of nitrate to nitrite, Voges-Proskauer reaction, degradation of tyrosine, phenylalanine, formation of indole, production of urease and egg-yolk lecithinase, and hydrolysis of hippurate were determined according to Gordon et al. [14].

The utilization of citrate and propionate was determined using the Simmon's medium [15]

The consumption of substrate by the new isolates was tested using a basal medium consisting of (grams per liter in distilled water): $(NH_4)_2HPO_4$ - 1; NaCl - 0.2; $MgSO_4$ - 0.2; yeast extract - 0.2; agar - 15. The hydrolysis of polysaccharides (carboxymethylcellulose, pectin, inulin) and the utilization of different carbohydrates as carbon sources (D-raffinose, maltose, D-galactose, cellobiose, D-fructose, L-rhamnose, sorbose, lactose, dulcitol, D-sorbitol, inositol) were tested. The substrates were added to a final concentration of 1% (w/v).

The casein and starch hydrolysis were tested by streak plate technique [13].

The gelatin liquefaction was determined by inoculating the bacilli into gelatin nutrient medium (gelatin 12g; distilled water 100ml) and incubation at 56 ^{0}C for 24 hours [12].

The lipolytic activity was determined by the hydrolysis of Tweens 20 (polyoxyethylene sorbitanmonolaurate), 40 (polyoxyethylene sorbitanmonopalmitate), 60 (polyoxyethylene sorbitanmonostearate), 80 (polyoxyethylene sorbitanmonooleate), tributirin, tricaprilin, sunflower oil [13].

The basal medium with inverted Durham tubes was used to test acid and gas production from D-(+)-glucose, L-(+)-arabinose, D-(+)-xylose and D-(-)-mannitol .

The resistance to lysozyme was studied by inoculating into a medium containing 10,000 units/ml enzyme.

Antibiotic Sensitivity

Sensitivity of the strains to antibiotics was tested using the enrichment-solid medium and sensitive discs. The following antibiotics were used: gentamicin, erythromycin, streptomycin, tetracycline, teramycin, rifampicin, levomycetin and penicillin G.

Enzymatic Activities

CGT-Ase Activity

The cyclodextrin glucanotransferase (CGTase) activity of the thermophilic cultures was revealed by plate express-method on solid nutrient medium containing stains [16].

The microscopic iodine test was used to screen the strains for CGTase activity [17]

Thin layer chromatography (TLC) of CGT-ase reaction products was done on Syluphol plates in *n*-butanol-acetic acid-water (4:3:3, v/v). The plates were treated by heating at 110-120°C and spraying with aniline phthalate solution for detection of sugars.

The quantitative determination of CGT-ase activity in culture broth was conducted by colorimetric method [18].

β-galactosidase Activity

The β-galactosidase activity was assayed by spectrophotometric measurement by Miller method [19]. One unit of the enzyme activity was defined as the amount releasing 1μmol of *o*-nitrophenol per min.

Protease Activity

The protease activity was examined by the modified method of Anson [20] with Na-caseinate as a substrate.

Amylase Activity

Amylase activity was routinely measured at 56^0C in 0.22 ml of the reaction mixture containing 0.2 ml of 1,0% (wt/vol) solution of soluble starch (from potato; Sigma) in 50mM Tris-HCl buffer (pH 7,2) and 0.02 ml of supernatant obtained by centrifugation (12000g, 15 min) of the cultural broth. The formed reducing sugar was measured by the Somogyi and Nelson method [21, 22]. One unit of enzymatic activity was defined as the amount of protein that produced 1 μmol of reducing sugar as glucose per min under the conditions of the assay.

Lipase Activity

Lipase activity was measured at 56^0C by alcalimetric titration [23] of 5 ml of the reaction mixture containing 1ml water with 0.05 ml Tween 20, 40, 60, tributirin, tricaprilin or sunflower oil, 3ml of 1/15 M phosphate buffer (pH 7,2), 0.2 mg bromtimolblue stain, 1ml of culture broth. One unit of enzymatic activity was defined as the amount of ml of 0,05 N KOH spent to titrate fatty acids.

Protein was determined according to Bradford [24] with albumin as the standard protein.

DNA Extraction and Purification

Genomic DNAs from bacterial strains were extracted and purified from 0.5 ml of each culture using the Genomic DNA Purification Kit (Fermentas).

PCR Amplification

Bacterial sequences were amplified in a *Mastercycler* thermal cycler (Eppendorf). Recombinant *Taq* DNA polymerase was used for PCR which was started with three primer sets: w001 + w002, w001 + w007, w001 + w012 (Table 1) [25]. Target DNA (0.2–0.4 μg), 0.2 mM of each primer, *Taq* polymerase 5 u/μl, 2 mM dNTP mix, 5 μl of 10 × PCR buffer, 25 mM $MgCl_2$ (Fermentas) were combined and adjusted to final volume of 50 μl with sterile double-distilled water. After an initial denaturation (95 ^{0}C, 4 min), 30 thermal cycles followed with the final denaturation set to 95 ^{o}C; 1 min; annealing at 50 ^{0}C 1 min; and primer extension at 72 ^{0}C 2 min. Amplificates were visualized by gel electrophoresis on 0.7 % agarose gels in 10×TBE buffer (89 mM Tris-borate, 2 mM EDTA) (Serva); samples were viewed by staining with ethidium bromide (Sigma) and photographed under UV light.

Table 1. Primers used in this study

Primer	Sequence
w001	5'-d(GTTTGATCMTGGCTC)-3'
w002	5'-d(GNTACCTTGTTACGACTT)-3'
w007	5'-d(CTCGTTGCGGGACTTAAC)-3'
w012	5'-d(TACGCATTTCACCKCTACA)-3'

Cloning and Sequencing

The cloning was performed by ligating the amplification products into the plasmid vector pUC57 DNA and using competent cells of *E. coli* XL1-Blue strain for transformation as described in the manufacturer's instructions *InsT/Aclone™ PCR Product Cloning Kit* (Fermentas). Double-stranded plasmid DNA sequencing was performed using *CycleReader™ Auto DNA Sequencing Kit* (Fermentas) and *Cy™5 labelled M13/pUC Sequencing Primer* (-46), 22-mer and Cy™5 labelled *M13/pUC Reverse Sequencing Primer* (-46), 24-mer (Pharmacia Biotech). Sequences were obtained by cycle sequencing and the chain termination reaction using an automate laser fluorescence sequencer *ALFexpressII* (Amersham Pharmacia Biotech). Sequences were aligned using CLUSTAL W program [26]. BLAST program [27] was used to look for sequence similarity.

Results and Discussion

The isolated strains have the following properties:

Morphological Properties

Strain T-17 colonies after 24 h growth on nutrient agar were small, circular in form, flat, colorless, transparent, homogeneous, with smooth shiny surface and even edges, and were easily peeled from agar. The colonies were 1 to 3 mm in diameter, became opaque after sporulation and later formed white smooth scurf.

On the same solid nutrient medium, strain T-18 cells formed small, circular, round colonies, with rough surface and undulate edges, which usually were attached strongly to agar. Mounds and lobes consisting largely of slime were often accumulated on the colonies. They were red pigmented on slants. Aged cultures were usually brown. The colonies were 2 to 3 mm in diameter.

T-17 and T-18 strains were straight rods. The cells were 0.8-1 μm wide and 3-3.8 μm long. They formed oval endospores. In case of T-17 strain the endospores were located terminally and were wider than the mother cells. Endospores of T-18 cells were sub-terminally located and did not swell the sporangium.

Physiological and Biochemical Properties and Antibiotic Sensitivity

The strains were characterized by the growth potential depending on sources of carbohydrates, NaCl concentrations, pH ranges and temperatures.

The optimal growth of T-17 cells occurred at 60^0C with the lower and upper growth limits at 35^0C and 75^0C, respectively. The growth temperature for the T-18 strain cells was at 15 - 70 ^{0}C, with the optimum growth in the range of 35-56^0C.

The pH range for growth of both strains was 6.5-9.0 with the optimum at pH 7.0-7.5.

T-17 cells were able to grow in the presence of 0-2% salt, but the maximum growth was observed in its absence. T-18 cells grew in a medium containing 0 to 7% NaCl with an optimum at 2%.

Both isolates were chemo-organoheterotrophs, Gram-positive, motile cells. Growth was aerobic, although the T-18 strain grew under anaerobic conditions, too. The catalase test was positive for both strains. The tests on determination of urease, phenylalanine deaminase, indole and hydrogen sulfide formation, tyrosine decomposition, assimilation of propionate, and formation of gas from both glucose and nitrate under anaerobic conditions were negative.

The T-17 cells produced acid from D-glucose, D-xylose, L-arabinose, but not from D-mannitol; did not produce acethylmethylcarbinol; did not reduce nitrate to nitrite; did not utilize citrate. The egg-yolk reaction was negative. The culture did not grow in the presence of 0.001% lisozyme or 0.02% sodium azide; it hydrolyzed Tweens 20, 40, and 80, inulin, carboxymethylcellulose, but not gelatin, Tween 60, pectin or hyppurate. Casein and starch were weakly hydrolyzed. Growth-supporting substrates were: D-raffinose, maltose, D-galactose, cellobiose, D-fructose, L-rhamnose, lactose, D-sorbitol, inositol. No or weak growth was observed on L-sorbose and dulcitol.

The strain T-18 produced acethylmethylcarbinol: the pH values of the Voges-Proskauer broth at the end of reaction were 5.4 to 5.6. Acid, but not gas, was produced from D-glucose, D-xylose, L-arabinose and D-mannitol; nitrate was reduced to nitrite; citrate was utilized; the egg-yolk reaction was positive. The cells grew in media containing 0.001% lisozyme or 0.02% sodium azide ; they hydrolyzed casein, gelatin, starch, Tween 20, 40, 60 and 80, ONPG, inulin, cellulose, but not pectin or hyppurate. All carbohydrates and alcohols tested as sources of carbon and energy were utilized.

Differential phenotypic and chemi-taxonomic characteristics of the two new isolates are shown in Table 2.

Table 2. Phenotypic characteristics of two isolates

Characteristics	T-17 / T-18
Cell size [μm]	
Width	0.8 / 1
Length	3.8 / 3
Cell form	Rods / rods
Motility	+ / +
Spore formation	+ / +
Endospore form	Oval / oval
Endospore location	Terminal / subterminal
Sporangium swell	+ / -
Catalase	+ / +
Anaerobic growth	- / +
Voges-Proskauer test	- / +
pH in V-P broth	- / 5.2 – 5.4
Acid from	
D-Glucose	+ / -
L-Arabinose	+ / +
D-Xylose	+ / +
D-Manitol	- / +
Gas from D-Glucose	- / -
Hydrolysis of	
Casein	± / +
Gelatin	- / +
Starch	+ / +
Tween 20; 40 and 80	+ / +
Hyppurate	- / -
ONPG	- / +
Inulin	+ / +
Pectin	- / -
Carboxymethylcellulose	± / ±
Utilization of	
Citrate	- / +
Propionate	- / -
Degradation of tyrosine	- / -
Deamination of phenylalanine	- / -
Egg-yolk lecithinase	- / +

Table 2. Phenotypic characteristics of two isolates (Continued)

Characteristics	T-17 / T-18
Nitrate reduction to nitrate	- / +
Formation of	
Indole	- / -
Dihydroxyacetone	- / -
Urease	- / -
Growth with lysozyme present	- / +
Growth in NaCl (%)	
0; 2	+ / +
3-7	- / +
> 7	- / -
Growth at pH	
5.5	- / -
6-8.5	+ / +
> 9	- / ±
Growth at T°C	
10	- / -
15-30	- / +
35-70	+ / +
75-80	± / -

Symbols: + positive; - negative; ± weakly positive

Both strains were resistant to penicillin G, teramycin, levomycitin, and rifampicin. Among the tested antibiotics, the strain T-18 displayed sensitivity to erythromycin, tetracycline and gentamycin, and the strain T-17 to erythromycin, streptomycin, and tetracycline (Table 3).

Table 3. Antibiotic sensitivity of newly isolated strains

Antibiotic	Strains	
	T-17	T-18
Streptomycin	s	r*
Penicillin G	r	r*
Tetracycline	s	s
Teramycin	r*	r*
Erythromycin	s	s
Levomycitin	r	r*
Rifampicin	r*	r*
Gentamycin	r	s

Symbols: r - resistant; s - sensitive; r* - resistant, but the growth was depressed

DNA Amplification and Restriction Analysis

The morphological and physiological characteristics, as well as the phylogenetic position determined by the16S rDNA data analyses, suggest that T-18 strain should be designated as a

Bacillus licheniformis. The Bergey's *Manual of Systematic Bacteriology* [28] describes *Bacillus licheniformis* as a species able to grow in 7% NaCL and at the maximum temperature of 55°C. Manachini et al. [29] reported that isolates belonging to *Bacillus licheniformis*, regardless of the sample origin, grew in 12% NaCl and were able to grow at up to 60 ^{O}C. These features allowed considering *Bacillus licheniformis* as a species including moderate halophilic and thermotolerant biotypes. Our results showed that cells of strain T-18 were able to grow at 70 ^{O}C and to utilize a wide spectrum of carbon sources. Thus, we inclined to consider this strain either as thermotolerant or as thermophilic one.

By the morphological and biochemical features, the strain T-17 belongs to the genus *Bacillus* and could be identified as a strain of *Bacillus stearothermophilus*. The sequence analysis of the 16S rDNA indicates that phylogenetically the newly isolated strain could be related rather to *Pseudomonas* than to *Bacillus*, though it differs from all known *Pseudomonas* species.

Enzymatic Activities

The capability to grow on media with various carbon sources (Table 4) was a key point for the selection of enzymatic activities to be studied.

Table 4. Substrate consumed as single carbon source

Carbon Source	Strains	
	T-17	T-18
Starch	++	+++
Inuline	+	++
Cellulose	++	+
Pektin	-	±
Casein	+++	+
Gelatin	-	+
Margarin	+	+
Butter	-	+
Palm oil	+++	+++
Olive oil	++	++
Sunflower oil	+++	+++

Table 4 indicated that one could expect activity of amylase, lipase, CGT-ase, protease and β-galactosidase. Therefore, activities of these enzymes were examined for the strains T-18 and T-17 of the thermophylic bacilli. The results are presented in Table 5.

It is obvious from the table that the studied strain of *Bacillus licheniformis* T-18 is a favorable source of thermophilic amylase, lipase, protease and galactosidase.

The specificity in the production of CDs types by extremophilic bacilli was reported by several investigators [30]. Particularly, biosynthesis of a mixture of α-, β- and γ-cyclodextrins is characteristic for thermophilic strains of bacilli . The data presented in this paper show that both T-17 and T-18 strains also produced a mixture of α-, β- and γ-cyclodextrins (CDs), but α-CD was dominating.

Although the rate of CGT-ase activity of both strains is lower than the maximum values reported in previous works, the new isolates could be useful for biotechnological investigation due to their ability to form CDs not only from starch, but also from inulin, cellulose and pectin.

Both isolated strains are conserved in the Collection of Extremophilic Bacilli of the Department of Microbiology and Microbial and Plant Biotechnology (Yerevan State University) and at the Georisk Scientific Research Company.

Table 5. Data of some enzymatic activities of isolated strains

Enzymatic activities	**T-17**	**T-18**
Amylase, μmol/min.mg protein	0,3	1,57
Lipase, substrates (ml 0,05 N KOH / 1ml culture broth):		
Tween 20	0,10	0,60
Tween 40	-	0,30
Tween 60	-	0,35
Tributirin	-	1,30
Tricaprilin	-	1,10
Sunflower oil	0,80	1,30
CGTase (U/ml)	231	291
Cyclodextrins	α / β / γ	α / β / γ
Protease (caseinase) (U/mg protein)	15,5	32,75
β-galactosidase (U/g dry cells)	-	555

Acknowledgments

The authors owe thanks to Dr. Barbara Nicolaus for discussion of the paper and constructive comments.

References

[1] B. Nicolaus, A. Panica, M.C. Manca, L. Lama, A. Gambacorta, T. Maugeri, C. Gugliandolo, D. Caccamo: *System. Appl. Microbiol.*, 23, N 3, 426 (2000).

[2] A.H. Varnam, M.G. Evans: *Environmental Microbiology*. Manson Publishing Ltd., London, 2000. 160 p.

[3] H.A. Panosyan, P.V. Tozalakyan, A.V. Gasparyan, Yu.G. Popov: *Biolog. J. Armenia,* 54 (1-2), 104 (2002).

[4] Y.-H. Lee: *Biolog. J. Armenia*, 53 (Special Issue: Cyclodextrins), 48 (2001).

[5] M. Byrne, D. Johnson: *Bioche. Soc. Trans.*, 2, 496 (1975).

[6] D. Narinesingh, T. Ngo, K. Laidler: *Can. J. Biochem.*, 53, 1061 (1975).

[7] M. Paine, R. Carbonell: *Biotechnol. Bioeng.*, 17, 617 (1975).
[8] H. Faulstich, A. Schafer, M. Weckauf-Bloching: *FEBS Lett.*, 48, 226 (1974).
[9] P. Bunting, K. Laiidler: *Biochem.*, 11, 4477 (1972).
[10] S. Khare, M. Gupta: *Biotechnol. Bioeng.*, 31, 829 (1988)
[11] T. Park, A. Hoffman: *J. Biomed. Mater. Res.*, 24, 21 (1990).
[12] L.G. Loginova, R.S. Golovacheva, L.A. Yegorova: *Life of microorganisms under high temperatures*. Nauka, Moscow, 1966. 240 p. (in Russian).
[13] N.S. Yegorov (Ed.): *Guide to practical training on microbiology*. Moscow University Publishing House, Moscow, 1983. 215 p. (in Russian).
[14] R.E. Gordon, W.C. Haynes, C.H.W. Pang: *The Genus Bacillus: Agricult. Handbook*. Washington D.C., 1973. 283p.
[15] Gerhard et al. (Eds): *Methods of General Bacteriology*. Mir, Moscow, 1983. 535 p. (in Russian).
[16] Z.G. Avagyan, A.A. Khachatryan, T.S. Davidyan, M.O. Adamyan: *Biolog. J. Armenia*, 53 (Special Issue: Cyclodextrins), 245 (2001).
[17] E.B. Tilden, C.S. Hudson: *J. Bacteriol.*, 43 (2), 527 (1942).
[18] W.S. Hale, L.C. Rawlins: *Cereal. Chem.*, 28 (1), 49 (1951).
[19] H.J. Miller: *Experiments in Molecular Genetics*. Mir, Moscow, 1976. 436 p. (in Russian).
[20] M.L. Anson: *J. Gen. Physiol.*, 22, 79 (1938).
[21] M. Somogyi: *J. Biol. Chem.*, 195 (1), 19 (1952).
[22] N.A. Nelson: *J. Biol. Chem.*, 153 (1), 375 (1944).
[23] L.B. Lobyreva, A.I. Marchenkova: *Microbiologia*, 48 (2), 208 (1979). (in Russian).
[24] M.M. Bradford: *Anal. Biochem.*, 72, 248 (1976).
[25] J.J. Godon, E. Zumstein, P. Dabert, F Habouzit, R. Moletta: *Appl. Environ. Microbiol.*, 63, 2802 (1997).
[26] *http://www.ebi.ac.uk/clustalw/index.html*
[27] *http://www.ebi.ac.uk/blast2/*
[28] J.G. Holt, N.R. Krieg, P.H.A. Sneath, J.T. StALEY et al. (Eds) *Bergey's Manual of Determinative Bacteriology*. Williams and Willkins, Baltimore, 1994. 547 p.
[29] P.L. Manachini, M.G. Fortina, L.Levati, C. Parini: *System. Appl. Microbiol.*, 21 (4), 520 (1998).
[30] V.H. Abelian: *Cyclodextrins: production and uses*. NAS of Armenia, Yerevan, 1998. 387p. (in Russian).

In: New Research on the Environment and Biotechnology
Editors: A. Kuznetsov and G. Zaikov, pp. 159-174
ISBN 1-60021-042-2

Chapter 17

CHARACTERIZATION OF THE METHANOGENIC COMMUNITY FROM THE HIGH-TEMPERATURE DAGANG OIL FIELD (P.R. CHINA)

N. M. Shestakova[1], A. A. Grigoryan[1], Qingxian Feng[2], Fangtian Ni[2], S. S. Belyaev[1], A. B. Poltaraus[3] and T. N. Nazina[1,*]

1 Winogradsky Institute of Microbiology, Russian Academy of Sciences, prospect 60-letiya Oktyabrya, 7/2, Moscow 117312, Russian Federation

2 Dagang Oilfield Company, 300280 Tianjin, P. R. China

3 Engelhardt Institute of Molecular Biology, Russian Academy of Sciences, Vavilova str., 32, Moscow 119991, Russian Federation

ABSTRACT

The number of methanogenic microorganisms and the rates of methanogenesis from 14CH3-COONa and NaH14CO3 were determined in formation waters from the high-temperature Dagang oil field. Bacterial production of methane was registered by a radioisotopic method with both substrates. Cells of methanogens were revealed by culture-based methods in media supplemented with H2+CO2 or acetate. Initial enrichments of methanogenic microorganisms were studied by molecular (16S rRNA genes) techniques. 102 unique clones containing 16S rRNAs archaea-specific inserts were identified. It was revealed that 101 clones belonged to the order *Methanobacteriales,* and 55 of these were related to *Methanothermobacter thermoautorophicum*; 44 clones were related to unculturable archaea from the same genus, and 1 – to *Thermococcus sibiricus.* The presence of phylotypes of the genus *Methanothermobacter* indicated predominance of hydrogenotrophic methanogens in the Dagang oil field. The absence of phylotypes of acetotrophic methanogens in the studied thermophilic community allowed us to suggest that acetate was degraded indirectly in syntrophic processes, rather than directly by aceticlastic methanogens.

---

* Author for correspondence: Tamara N. Nazina. Tel: + 7 095 1350341. Fax: +7 0951356530. E-mail: nazina@inmi.host.ru

Keywords: subsurface ecosystem, thermophiles, methanogenesis, sulfate reduction, 16S rRNA, methanogenic archaea

INTRODUCTION

During the last decades, the microflora of subterranean ecosystems have been attracting attention of many researchers. Cultural, molecular, and biogeochemical techniques were used in order to investigate microbial biodiversity of the oil fields. These methods were usually used separately, providing therefore no possibility to decipher the structure of microbial community and to assess its geochemical activity. Using traditional, selective media-based techniques for the analysis of microbial communities enabled isolation and enumeration of culturable microbial forms only. For instance, in different oil horizons, anaerobic microorganisms were found that reduced sulfate, thiosulfate, elemental sulfur, manganese, and iron, as well as fermentative organotrophs, acetogens, and methanogens [see Magot et al., 2000, for review; Orphan et al., 2000, 2003; Bonch-Osmolovskaya et al., 2003]. In spite of the growing number of publications dealing with description of microorganisms from oil layers, enumeration of bacteria by molecular techniques or by direct microscopic count indicates that culturable organisms constitute only a minor fraction of microbial communities [Onstott et al., 1998, Bonch-Osmolovskaya et al., 2003; Orphan et al., 2000, 2003].

Methanogenic microorganisms of the oil layers are the least known group of prokaryotes. The presence of methanogens in oil layers was initially demonstrated by S.I. Kuznetsov [1950]. The first pure cultures of methanogens from oil layers were obtained by S.S. Belyaev and coauthors [1983, 1986; Obraztsova et al., 1987a, b; Davydova-Charakhch'yan et al., 1992; Davidova et al., 1997].

Methanogens performing hydrogen oxidation with concomitant reduction of carbon dioxide to methane are common in oil layers with temperatures 20 to 80 °C [Magot et al., 2000]. Mesophilic hydrogen-utilizing methanogens belong to the species *Methanobacterium bryantii, M. ivanovii, 'Methanoplanus petrolearius'* [Belyaev at al., 1983; Davydova-Charakhch'yan et al., 1992; Ollivier et al., 1998]. Although hyperthermophilic methanogens growing at 110 °C are known [Kurr et al., 1991], such organisms have not yet been found in oil fields. Lithotrophic methanogens belonging to the species *Methanococcus thermolithotrophicus* and to the different species of the genus *Methanothermobacter* (previously, the genus *Methanobacterium*) - *M. thermautotrophicus (M. thermoalcaliphilum,)* and *M. thermoaggregans* are responsible for methane production in high-temperature oil reservoirs of the USA, North Sea, and Western Siberia [Ng et al., 1989; Davydova-Charakhch'yan et al., 1992; Nilsen and Torsvik, 1996; Magot et al., 2000; Kotelnikova et al., 1993].

Organotrophic methanogens utilizing acetate, ethanol, and methylated amines, and often capable of growth also on H_2+CO_2, have been isolated from the oil fields with temperature not exceeding than 60 °C [Davidova et al., 1997; Obraztsova et al., 1987a,b]. Organotrophic mesophilic methanogens belong to the genera *Methanosarcina* (*M. mazei, M. siciliae*), *'Methanohalophilus' ('Methanohalophilus euhalobius' =Methanococcoides euhalobius*), and *Methanocalculus* (*Methanocalculus halotolerans*) [Davidova et al., 1997; Obraztsova et al., 1987a,b; Ni et al., 1994; Ollivier et al., 1998]. Mesophilic filamentous microorganisms similar to *Methanosaeta* (*Methanothrix*) were observed in methanogenic enrichments from the oil fields of Apsheron Peninsula [Nazina, Rozanova, 1980; Nazina, 1984].

Thermophilic syntrophic acetate decomposition with methane production was first described by Davydova-Charakhch'yan and coauthors [1992]. In the high-temperature Mykhpay oil field, acetate was decomposed by a binary association with a lithotrophic methanogen *M. thermautotrophicus* (=*M. thermoalcaliphilum*) as a terminal component [Davydova-Charakhch'yan et al., 1992; Kotelnikova et al., 1993]. The same species *M. thermautotrophicus* was found in the Samotlor oil field; acetoclastic methanogens growing at 60 and 80 °C were not found [Bonch-Osmolovskaya et al., 2003]. No hyperthermophilic acetate-utilizing methanogen is presently known.

However, methane production from $NaH^{14}CO_3$ *and* $^{14}CH_3COONa$ *detected by radioisotopic methods in the oil fields with 60 to 80 °C indicated the presence of thermophilic prokaryotes performing these processes in the oil layers [Belyaev, Borzenkov, 1993; Nazina et al., 1995, 2000; Rozanova et al., 1995, Bonch-Osmolovskaya et al., 2003].*

Recently, molecular biological techniques were applied for analysis of microbial communities in the oil fields [Voordouw et al., 1996; Orphan et al., 2000, 2003; Leu et al., 1998, Bonch-Osmolovskaya et al., 2003). Analysis of microbial communities based on nucleic acids extraction from natural samples with subsequent amplification of 16S rRNA gene fragments by means of polymerase chain reaction (PCR) and thorough analysis of these fragments by molecular genetic techniques provides the most comprehensive picture of the composition of a microbial community. This approach is widely used for analysis of microbial communities of hot springs, marine hydrotherms, and terrestrial habitats. Similar investigation was performed by Orphan and coauthors for the Californian oil layers with temperatures 50 to 125 °C [Orphan et al., 2000, 2003]. In formation waters, 16S rRNA genes of thermophilic organisms belonging to the genera *Thermococcus*, *Petrotoga*, *Thermoanaerobacter*; *Methanothermobacter (Methanobacterium)*, *Methanococcus*, and *Methanoculleus* were found, together with those of mesophilic bacteria and of new archaea and bacteria. Autotrophic methanogens of the genera *Methanothermobacter* (*Methanobacterium,* growth at 60-70 °C), *Methanococcus* (growth at 60-80 °C) and *Methanoculleus* (60-70 °C) were recovered from Californian oil reservoirs using culture-based methods. In the Monterrey Formation, a 16S rRNA sequence of a new acetoclastic methanogen forming a deep branch in the order *Methanosarcinales* was found also [Orphan et al., 2000].

The objective of our study was to estimate simultaneously the number, the metabolic activity and phylogenetic diversity of methanogens inhabiting the high-temperature Dagang oil field. This included the following particular goals: (1) Enumeration of representatives of the main metabolic groups by cultural methods; (2) Estimation of the rates of anaerobic processes of sulfate reduction and methanogenesis in water samples by radioisotopic methods; (3) Characterization of the methanogens' diversity by 16S rRNA molecular analysis.

2. Materials and Methods

2.1. Site Description and Media Composition

Characterization of the Dagang Oil Field

The distribution of microorganisms in the Kongdian bed of the Dagang oil field located in the Hebei province (P.R. China) was studied. The sandstone oil-bearing horizons studied had a temperature of 59 °C, an average porosity of 33%, a permeability of 1.878 μm^2. The oil of the Dagang oil field has density about 0.900 g/cm^3, containing 53% saturated hydrocarbons, 20% aromatic compounds, and 21.15% resins+asphalthenes and has a viscosity 73 mPa x s at 50 °C. The gas contains methane (95-98%), its higher homologues (0.8-1.8%), N_2 (0.5-3.3%) and CO_2 (0.06-0.77%). The formation water of the hydrocarbonate-sodium type has low salinity (5612 mg/1).

Samples of injection and formation waters were taken in June and December 2000 and in June 2002. Bacteria were isolated from the formation water that was effluent from depths of 1206-1435 m through production wells and through injection wells operating in a backflow mode. Samples of injection water and production fluids were collected directly from wellheads into sterile bottles and analyzed in the laboratory within 4-6 h after sampling for bacterial numbers and rates of microbial processes. The fluids were used also for concentration of microbial biomass for molecular analysis. The pH, temperature, sulfide and oxygen concentration of formation waters were measured at the time of sampling. Samples were stored at 6 °C until completion of all chemical analyses.

Enumeration of Bacteria

A technique for cultivation of strictly anaerobic microorganisms [Hungate, 1969] was applied for the cultivation and enumeration of microorgansims. Anaerobic bacteria were cultivated in tubes with pure argon in the gas phase; an exception was the medium with H_2/CO_2 for methanogens. The numbers of cultivable anaerobic bacteria were estimated by serial decimal dilutions of water samples in tubes with various enrichment media designed to promote growth of specific functional groups. The highest dilution in which growth occurred was assumed to approximate the inverse log of the population size, as determined by the most probable number (MPN) procedure [Koch, 1994]. Enrichment media were degassed by boiling under a stream of O_2-free Ar (for fermentative bacteria) or prereduced (for sulfate-reducing bacteria and methanogens) and were adjusted to pH 7.0-7.2. Anaerobic organotrophic bacteria with a fermentative type of metabolism were determined by measuring the increase in H_2 in medium with peptone (4 g l^{-1}) and glucose (10 g l^{-1}) [Postgate, 1984]. Sulfate-reducing bacteria were analysed by measuring the increase in H_2S in medium B containing sodium lactate (3.5 g l^{-1}) and reduced with $Na_2S^x9H_20$ (0.2 g l^{-1}) [Postgate, 1984]. Methanogens were determined by measuring the increase in CH_4 in media [Zeikus et al., 1975] with acetate (2.2 g l^{-1}) or H_2/CO_2 (80:20 % v/v), supplemented with microelements and yeast extract (1 g l^{-1}) [Wolin *et al.,* 1963].

Aerobic bacteria were cultivated in Hungate tubes with air as the gas phase. Viable aerobic hydrocarbon-oxidizing bacteria were also enumerated by serial decimal dilutions of water samples in P medium with a mixture of saturated C_{10}-C_{22} hydrocarbons (1 % v/v) [Nazina *et al.,* 1985].

All media were inoculated with samples of formation water using syringes. Tubes were incubated at 60 °C for at least 30 days or until no further growth was observed. All tubes were examined by phase-contrast microscopy and with using analytical methods for determination of the metabolic products of microorganisms – H_2S, CH_4, CO_2, H_2.

2.2. Analytical Methods

The chemical composition of the formation water was determined according to recomendations [Reznikov et al., 1971]. Hydrogen sulfide was determined with dimethyl-p-phenylenediamine by a modified colorimetric method of Pachmayr [Truper, Schlegel, 1964].

Molecular hydrogen was analyzed with a model 3700 gas chromatograph ("Chromtograph", Moscow, Russia). A column (2 m x 2 mm ID), packed with the 5A molecular sieve, mesh 60/80, was used for analyses. The carrier gas was argon. The thermal conductivity detector and injector were kept at 70 and 40^0C, respectively. Methane was determined with the same chromatograph equipped with a flame ionization detector and a SOVKOL prepacked column (1 m x 3 mm ID). The carrier gas was argon. The detector and injector were kept at 150 and 140°C, respectively. The standard gas mixtures of H_2 and CH_4 produced by Balashikha Oxygen Plant (Balashikha, Russia) were used for calibration of gas chromatograph.

Concentration of lower fatty acids was determined in samples fixed with saturated KOH (2 ml/50 ml of sample). In the laboratory, 0.98 ml of each sample was acidified with 0.02 ml of 25% HCl and analyzed on a model 3700 chromatograph ("Chromatograph", Moscow, Russia) equipped with a flame ionization detector. The column (2 m x 3 mm ID) was packed with Porapak Q (100/120 mesh). The temperature of injector was 180°C. A mixture of N_2 (95%) and CO_2 (5%) was used as the carrier gas. The standard solutions of lower fatty acids produced by Chemical department of the Lomonosov Moscow State University (Moscow, Russia) were used for calibration of the chromatograph.

2.3. Molecular Methods of Cloning and Sequencing 16S Rrna Genes

Nucleic Acid Extraction

Active methanogenic enrichments obtained on media with H_2+CO_2 and acetate (from production wells 1017-7, 1008-1, 1017-2, 1017-4 and 1094-1) were pooled together and used for DNA extraction. The cells from 100 ml of active cultures were harvested by centrifugation at 5,000 rpm for 30 min. The cells were resuspended in 1 ml of 0.9% NaCl, transferred to a 1.5-ml Eppendorf tube, and centrifuged at 14,000 rpm for 5 min. The supernatant was decanted, and the pellet was used for DNA extraction using «Diatom™DNAprep» kit (IBCh RAS, Moscow, Russia) according to the manufacturer's instructions. The DNA pellet was resuspended in 100 µl of double-distilled water (ddH_2O) and stored at -20 °C.

Amplification of 16S rRNA genes

The isolated genomic DNA was used as a template in the polymerase chain reaction with universal archaeal oligonucleotide primers. The reaction mixture (50 μl) contained 10-50 ng of the DNA template; standard concentrations of dNTP (200 μM); primers, 0.1 μM; the thermostable Taq-DNA-polymerase, 1,25 U (Perkin-Elmer); $MgCl_2$, 2 mM (required for the functioning of the latter enzyme), and the buffer containing 10 mM Tris-HCl (pH 8.3), 50mM KCl, and 0.1 vol % Tween 20.

Archaeal 16S rRNA genes were amplified by performing PCR with the A109f forward primer - 5'-ACG/TGCTCAGTAACACGT-3' corresponding to positions 109 to 125 of the *Escherichia coli* 16S rRNA gene [Brunk et al., 1996; DeLong, 1992] and the A1041r reverse primer - 5'-GGCCATGCACCWCCTCTC-3' corresponding to positions 1041 to 1058 [Kolganova et al., 2002].

DNA amplification was performed with a model iCycler *"BioRad"* by using the following program for PCR amplification with archaeal primers: 3 min of polymerase activation at 94°C, followed by 35 cycles consisting of denaturation (0.5 min at 94°C), primer annealing (0.5 min at 50°C), and chain extension (1 min at 72°C), a final extension step at 72°C for 7 min. The amplified products were separated in 0.8 % agarose gel that was stained with ethidium bromide and were visualized with UV light and excised together with agarose, which was then melted at 70°C for 15–20 min. PCR products were purified by using V-gene DNA Gel Extraction Kit (V-gene Biotechnology Limited, Hangzhou 310022, China).

Cloning of 16S rRNA Genes

PCR products (about 1000 nucleotides of archaeal 16S rRNA genes) were purified by extraction from agarose gel and cloned into plasmid vector pGEM-T (Promega) by using a TA cloning vector kit according to the manufacturer's instructions. Recombinant plasmids were introduced into *E. coli* Z85 competent cells by electroporation. White clones were screened for the presence of insert-containing plasmids by PCR using vector-specific primers (SP6 and T7). A total of 102 white colonies from methanogenic enrichments were selected and screened for the archaeal (A1) library. Plasmid DNA from unique clones was purified by the alkaline lysis method [Birnboim, 1983] with some modifications and used for sequencing reactions. Alternatively, all white colonies originating from the transformation experiments were picked and used as template for reamplification of the 16S rRNA inserts. PCR products were purified by ethanol precipitation and used for DNA sequencing.

Sequencing of Cloned 16S rRNA Genes Fragments

The nucleotide sequence of the 16S rRNA gene was determined by enzymatic sequencing on an automatic Applied Biosystems DNA Sequencer (ABI 3100 Avant Genetic Analyser) and using Amersham kit (Dyenamic Terminator Cycle Sequencing Ready Reaction kit) according to the recommendations of the manufacturers. The archaeal 16S rDNA fragments were sequenced using A109f primer. The sequences obtained by manual sequencing were assembled using the SeqMan program of the DNAStar software package (DNASTAR Inc., Madison, Wis.).

Phylogenetic Analysis of Sequences

The nucleotide sequences of the 16S rRNA-encoding gene from clones were submitted to GenBank for preliminary analysis using the BLAST software of the Ribosomal Database Project to determine putative close relatives [Maidak *et al.,* 1992]. Sequences were aligned manually against the nearest neighbor. Mutual evolutionary distances (expressed as estimated changes per 100 nucleotides) were computed by use of the correction of Jukes and Cantor [1969].

2.4. Radioisotope Methods

The rates of sulfate reduction and methanogenesis were determined by radioisotope methods using corresponding labeled compounds: $Na_2{}^{35}SO_4$, $^{14}CH_3$-COONa, and $NaH^{14}CO_3$ [Ivanov, Terebkova, 1959; Belyaev, Ivanov, 1975; Laurinavichus, Belyaev, 1978].

3. Results

3.1. Microbiological Characteristics of Formation Waters of the Dagang Oil Field

Chemical and microbiological composition of formation water of the Dagang oil field was studied in a period from 2000 to 2002. Formation waters from 22 monitor production wells were used in this study. Oil-bearing horizons of the Dagang oil field had a temperature 59 °C. The Kongdian oil bed had been exploited with water flooding to maintain the formation pressure. Coproduced water of the Kongdian oil bed was separated from oil for 2-5 hours and was used again for injection. Injection water had a temperature 35-50 °C and contained about 0.8-1.6 mg of dissolved oxygen per 1 l.

In 2001, a pilot test of a technology of the microbial enhancement of oil recovery (MEOR) was started on the Kongdian bed of the Dagang oil field. According to the procedure used on the Kongdian oil bed, oxygen (as water-air mixture) and mineral nitrogen and phosphorus salts were injected into the oil field through 8-12 injection wells. The microbiology and geochemistry of this subsurface ecosystem were characterized both before and after nutrient addition.

In 2000, baseline analyses for 9 and 3 months prior to nutrient injection showed that injection waters contained all the groups of microorganisms under investigation. The components of the microbial community of the injection water were mesophilic and thermophilic aerobic hydrocarbon-oxidizing bacteria and anaerobic fermentative, sulfate-reducing and methanogenic bacteria.

Microbiological analyses of water from the production wells revealed the presence of aerobic and anaerobic thermophilic microorganisms. A small number of aerobic organotrophs growing on C_{10}-C_{22} saturated hydrocarbons (0-single cells and up to 10^2 cells/ml in 3 of 22 production wells) was obtained. Anaerobic fermentative (10^2-10^5 cells/ml), sulfate-reducing (0-$6.0 \cdot 10^3$ cells/ml) and methanogenic microorganisms dominated in the oil field.

Methanogenic archaea were presented mainly by rods growing in media with H_2+CO_2 (0-6.0·10^3 cells/ml) and with acetate (0-10^2 cells/ml).

The addition of the nutrient solution into the oil field favored the growth of aerobic and anaerobic microorganisms. From the near-bottom zone of the injection well 1098 operated in a back-flush mode, formation water was obtained (sample of water 24 m^3) with residual oil, black sediment, and viscous and slimy biofilms. The number of all studied groups of microorganisms here was 10-100 times higher than in the zone of production wells.

In water samples from production wells, the number of aerobic hydrocarbon-oxidizing bacteria did not exceed 10^2-10^3 cells/ml (Table 1).

Table 1. Number of aerobic and anaerobic microorganisms in formation waters of the Dagang oil field (cells · ml^{-1})

Well number	Hydrocarbonoxidizing	Fermentative**	Sulfate-reducing	Methanogenic	
	C_{10}-C_{20} hydrocarbons	Peptone + Glucose	Lactate	H_2+CO_2	Acetate
1098* – 8 m^3	10^3	10^7	10^5	10^6	10^5
1098* – 24 m^3	10^5	10^7	10^3	10^4	10^4
Production wells:					
01	Single cells	10^5	10	10^2	10
63	0	10^5	10^3	10^3	10
1002-1	0	10^7	10^2	10^4	10^3
1008	0	10^5	10^2	10	10
1008-1	10^2	$\geq 10^7$	10^2	10^3	10^2
1012	10^2	$\geq 10^5$	10	10^3	10
1012-1	10^3	$\geq 10^5$	10^3	10^3	10^2
1015-1	0	10^8	10^3	10^3	10^3
1017	0	$\geq 10^4$	10	10^3	Nd
1017-3	0	10^8	10	10^4	10^2
1017-5	0	10^4	10	10^2	Single cells
1017-7	Single cells	10^7	10^2	10^4	10^2
1032	0	$\geq 10^6$	Single cells	10^2	10
1032-1	Single cells	$\geq 10^4$	Single cells	10^3	10
1050-1	0	$\geq 10^4$	10^4	10^2	10
1050-2	Single cells	10^8	Single cells	10^2	10^2
1050-3	Single cells	10^6	10^4	10^3	10
1092	Single cells	10^7	10^2	10^3	10^3
1094	0	10^6	10^2	10^3	10
1094-1	Single cells	10^5	$\geq 10^5$	$\geq 10^4$	$\geq 10^4$

*, Water samples from the near-bottom zone of injection well;
**, The number of fermentative bacteria as the result of microscopic studies

The number of fermentative bacteria increased up to 10^6-10^7 cells/ml. An increase in the population of fermentative bacteria by several orders of magnitude reflected the total enrichment of subsurface microbial community. Fermentative bacteria exhibited close metabolic interactions with other anaerobic bacteria – sulfate-reducers and methanogens. In cultures of fermentative bacteria, we registered production of hydrogen sulfide (wells 63, 1050-3 and 1092) and methane (wells 1017-4 and 63). Anaerobic thermophilic enrichment cultures producing methane in media with pyruvate, ethanol, or benzoate were obtained from the oil field. We believe that methane was produced as a result of syntrophic growth of microorganisms.

In the course of the nutrient injection, the number of sulfate-reducing bacteria did not change essentially and increased only in formation waters of six wells (Table 1). The number of methane-producing archaea reached 10^4 cells/ml (wells 63, 1002-1, 1012, 1017-3, 1017-7, 1094-1). As a rule, the number of methanogens in the medium with H_2+CO_2 was 10 times higher than in the medium with acetate. Methanogenic enrichments were studied by fluorescent microscopy with excitation wavelengths close to 420 nm (blue-violet). In both nutrient media - with H_2+CO_2 or acetate, autofluorescence was revealed in microorganisms represented by rod-shaped cells of various lengths.

3.2. Chemical Characteristics of Formation Waters and Gas of the Dagang Oil Field

Chemical analyses of formation waters prior to nutrient injection showed that it belonged to the hydrocarbonate-sodium type. The water was slightly alkaline, with pH from 7.2 to 7.6, exhibited low total salinity ranging from 5.8 to 6.7 g/l, a relatively high content of hydrocarbonate (419-698 mg l^{-1}). Content of acetate (mainly less than 5 mg l^{-1} and only in the production well, 63 - 60 mg l^{-1}) and SO_4^{2-} (0-55 mg l^{-1}) was low; H_2S was revealed in samples from the near-bottom zone of injection wells in a small quantity (less than 1 mg l^{-1}) (data not presented).

The addition of nutrient solution into the near-bottom zone of injection wells favored the biotransformation of oil in this zone. Both HCO_3^- and lower fatty acids are usual products of microbial oil oxidation. Products of oil oxidation appeared in formation waters; they migrated from the zone of injection wells in the direction of production wells and stimulated the growth of anaerobic microorganisms. In June 2002, in water of many production wells, an increase in the concentration of soluble carbonates as compared to the background value (from 0.4-0.6 to 0.7-1.052 g l^{-1}) was recorded (Table 2). In 6 of 22 production wells total concentrations of acetate + formate + iso-butyrate exceeded 150 mg l^{-1}. As a rule, propionate was absent. The highest concentration of acetate reached 86.6-160.7 mg l^{-1} as compared to 5-60 mg l^{-1} before the nutrient flooding.

The change in the gas chemical composition during the nutrient flooding of the oil field was not high. A small increase in carbon dioxide (0.5-1.7%) and methane (3-5%) was observed in wells 1015-1, 1050-3, 1017-2 and 1094-1.

3.3. Estimation of the Rates of Sulfate Reduction and Methanogenesis in Formation Waters

Before the nutrient injection, thermophilic microbial processes of sulfate reduction and methanogenesis were most pronounced in the near-bottom zone of injection wells of the Dagang oil field. The rate of sulfate reduction there did not exceed 19.90 μg of S^{2-} l^{-1} day^{-1}. The rates of methanogenesis from labeled bicarbonate and acetate reached 1.505 and 14.73 μg of CH_4 l^{-1} day^{-1}, respectively. In the zone of production wells, the rate of sulfate reduction was in the range of 0.002 to 18.9 μg of S^{2-} l^{-1} day^{-1}, a rather high value, taking into account the low sulfate concentration in formation waters. The rates of methanogenesis from bicarbonate and acetate were in the range of 0-1.595 and 0-5.61 μg of CH_4 l^{-1} day^{-1}, respectively.

Table 2. Chemical composition of formation water and rates of thermophilic microbial processes in the Dagang oil field (June 2002)

Well number	Total salinity, mg · l^{-1}	Acetate, mg · l^{-1}	Formate, mg · l^{-1}	i-Butyrate, mg · l^{-1}	SO_4^{2-}, mg · l^{-1}	HCO_3^-, mg · l^{-1}	Methanogenesis Rate, µg of CH_4 · l^{-1} · day^{-1}		Sulfate Reduction Rate, µg of S^{2-} · l^{-1} · day^{-1}
							From $NaH^{14}CO_3$	From $^{14}CH_3COONa$	
1098* – 8 m^3	6314	16	Nd	Nd	12	534	0.693	228.475	291.296
1098* – 24 m^3	6279	30	Nd	Nd	12	488	0.635	40.200	20.123
Production wells:									
01	6283	22.6	15.6	0	30	412	0.048	0.052	5.179
63	6952	4.8	6.6	10	41	549	0.253	0.014	297.052
1002-1	6827	74.4	67.4	98.2	0	747	0.151	20.55	0.009
1008	7064	1.0	2.0	1.2	12	610	0	0.003	0.187
1008-1	6616	160.7	35.8	52.2	12	747	0.569	39.680	215.629
1012	6852	70.8	23.6	31.6	47	610	0	1.184	5.49
1012-1	6897	70.8	10.8	17.4	30	579	0.039	14.054	0.574
1015-1	6544	52.2	39.2	95.4	0	564	0	0.716	0.012
1017	6553	11.6	3.0	23.0	0	488	0.167	0.04	0
1017-3	6786	86.6	28.0	67.8	0	732	0.873	1.768	0.02
1017-4	6522	64.2	17.2	19.2	24	503	0.161	42.255	0.795
1017-5	7005	119.9	27.0	80.4	35	625	0.231	6.969	12.029
17-7	7185	0.5	1.0	22.0	6	808	0.378	0.162	1.445
1032	6737	5	4.6	0	0	457	0.04	0.006	0
1032-1	6003	7.4	1.6	17.8	12	579	0	0.640	0.044
1050-1	6320	70.6	14.8	26.6	53	457	0.559	0.236	16.306
1050-2	6688	55.6	22.2	37.6	0	457	0.03	0.192	0
1050-3	6938	2.8	8.4	14.0	0	640	0.2	0.018	0
1092	6459	65.5	61.4	62.0	47	579	15.668	0.169	159.698
1094	6767	26	3.4	13.4	35	640	1.04	0.175	0.217
1094-1	6080	53.4	45.6	58.6	47	1.052	12.066	1143.479	29.970

*, Water samples from the near-bottom zone of injection well.

In the course of nutrient injection, the rate of sulfate reduction in major portion of the stratum was in a range of background values, but it increased significantly in the near bottom zone of injection well 1098, where the microbial community was mostly induced, and in wells 63, 1008-1, and 1092 (159.69-297.05 μg of S^{2-} l^{-1} day^{-1}). The rate of methanogenesis also increased in the near-bottom zone of injection well (Table 2) and reached 229.17 μg of CH_4 l^{-1} day^{-1}. The most significant increase was found for methanogenesis from acetate. It can be explained by appearance of high quantities of lower fatty acids as a result of biotransformation of oil. In most of the production wells zones, the rate of methanogenesis from $NaH^{14}CO_3$ was in a range of 0-15.668, and from $^{14}CH_3$-COONa, in a range of 0.006-42.255 μg of CH_4 l^{-1} day^{-1}. In the production well 1094-1, methanogenesis rate from acetate became exclusively high and reached 1143.479 μg of CH_4 l^{-1} day^{-1} (this finding correlated with high numbers of methanogens and a high level of additional oil recovered from this well, data not presented). Thus, in formation waters of the Dagang oil field methane was produced from both labeled compounds - $NaH^{14}CO_3$ and $^{14}CH_3$-COONa.

3.4. Characterization of Methanogenic Enrichments of the Dagang Oil Field by Means of 16S rRNA Molecular Analysis

As it was shown above, in the Dagang oil field we registered methanogenic microorganisms and the processes of methanogenesis both on H_2+CO_2 and on acetate. In June 2003, samples of formation water with the highest rates of methanogenesis and the highest numbers of methanogenic bacteria were selected for 16S rRNA-based molecular studies. Active methanogenic enrichments from production wells (1017-7, 1008-1, 1017-2, 1017-4 and 1094-1) obtained in media with H_2+CO_2 or acetate were combined and used for DNA extraction. Archaeal 16S rRNA genes were amplified. 16S rRNA genes of archaea (about 1000 nucleotides) were cloned into plasmid vector pGEM-T. One 16S rDNA clone library (A1) was generated from methanogenic enrichments of the Dagang oil field. The archaeal A1 library contained 102 clones.

In total, 102 unique clones were sequenced from archaeal library. Affiliation of these clones from the methanogenic enrichments is shown in Table 3. Archaeal phylotypes showed small sequence diversity and were represented by members of the order *Methanobacteriales* (101 clones) and the order *Thermococcales* (one clone). Sequence similarity of 16S rRNA of clone A116 and *Thermococcus sibiricus* was high (99.1%) and allowed to identify this clone as a representative of this species. Of the 101 clones of the order *Methanobacteriales*, 55 were highly similar (97-100%) to *Methanothermobacter thermoautotrophicus*, widespread in other high-temperature oil fields. 44 clones were also close to an uncultured representative of the genus *Methanothermobacter*. Two clones, A63 and A104, were phylogenetically close to clone E21A1 from an enrichment culture obtained from the Monterey production fluids [Orphan et al., 2000]. Both clone E21A1 and clones A63 and A104 from the Dagang oil field were phylogenetically close to *M. thermautotrophicus*. The main substrates of methanogenesis for the representatives of the order *Methanobacteriales* are known to be H_2+CO_2. Thus, we did not find phylotypes of acetate-utilizing methanogens in the studied thermophilic community.

Table 3. Phylogenetic diversity of 16S rRNA genes in methanogenic cultures isolated from the Dagang oil field

Dominion/order	Phylogenetically closed sequence in RDP	Number of clones	Nucleotide sequence similarity, %
Archaea			
Methanobacteriales	*Methanothermobacter thermautotrophicus*	55	97.3 – 100
	Methanothermobacter spp.	44	97.4 – 99.5
	Clone E21A1 (*M. thermautotrophicus*)	2	98.6
Thermococcales	*Thermococcus sibiricus*	1	99.1

4. Discussion

The diversity and activity of the microbial community of the high-temperature water-flooded Dagang oil field were studied using several approaches. The number of thermophilic microorganisms of the main metabolic groups (fermentative, sulfate-reducing, methanogens) was estimated by cultural methods, the rates of methanogenesis and sulfate reduction were determined by radioisotopic methods. Phylogenetic diversity of methanogens in methanogenic enrichments was studied by 16S rRNA molecular analysis. This work was of special importance because the microbial community was analyzed both prior to and in the course of application of a biotechnology, based on injection of oxygen (air) and nitrogen and phosphorus salts to activate the microorganisms inhabiting the oil field.

Prior to nutrient flooding, the microbial community of the Dagang oil field included fermentative, sulfate-reducing and methanogenic microorganisms; their numbers did not exceed 10^5, $6.0x10^3$ and $6.0x\ 10^3$ cells/ml, respectively. Aerobic oil-oxidizing bacteria ($0\text{-}10^2$ cells/ml) were also present in formation waters. Both bacterial numbers and the rates of the terminal microbial processes of oil biodegradation were low and increased in the near-bottom zone of injection wells, where a small amount of oxygen dissolved in injection water penetrated. The rates of sulfate reduction and methanogenesis were similar to those in the Mykhpaiskoe, the Talinskoe, the Samotlor (all located in Western Siberia, Russia) and the Liaohe (P.R. China) high-temperature oil fields, also exploited with water flooding [Nazina et al., 1995, 2000; Bonch-Osmolovskaya et al., 2003).

Subsurface ecosystems are known to be limited on the quantity of electron acceptors and nitrogen and phosphorus nutrients. The injection of oxygen and of nitrogen and phosphorus salts significantly stimulated the process of oil biotransformation in the oil field. The cell number of thermophilic, oil-oxidizing, fermentative, sulfate-reducing and methanogenic bacteria increased $10\text{-}10^4$-fold. The rates of methanogenesis and sulfate reduction increased in the near-bottom zone of injection wells and in the zone of production wells. Microbial oil transformation was accompanied by accumulation of bicarbonate ions and lower fatty acids in formation waters and of CH_4 and CO_2 in gas.

The increase in numbers of methanogenic organisms together with the high rate of methanogenesis in the near bottom-zone of injection wells induced by nutrient injection strongly suggests that methanogens were active during the hiatus in aerobic oil oxidizing activity that occurred in the oil field. The evidence of methanogenesis along with the substantial increase in numbers of aerobic bacteria during injection of oxygen and of nitrogen and phosphorus salts suggests that this was a result of oxygen consumption for oil oxidation by aerobic bacteria. In addition, the content of oxygen decreased to the concentration sufficient to permit a level of anaerobiosis commensurate with methanogenesis. Methanogenic activity could have occurred in microhabitats (micropores) in the sediment matrix. Sulfate-reducing bacteria (also strict anaerobes) were present together with aerobes in the water samples from the near bottom-zone of injection well. These results indicate that sulfate-reducing and methanogenic microorganisms are confined to the zone of oil oxidation. In the zone of production wells, the highest rates of sulfate reduction and methanogenesis were also observed when oil biodegradation products (bicarbonate ions and lower fatty acids) were present in the water.

In the Dagang oil field, we registered the methanogenic microorganisms and the methanogenesis processes both on H_2+CO_2 and on acetate, as it was demonstrated in other high-temperature petroleum reservoirs [Belyaev, Borzenkov, 1993; Rozanova et al., 1995; Nazina et al, 1995, 2000; Bonch-Osmolovskaya et al., 2003]. Unfortunately, not all important biogeochemical processes that are mediated by bacteria provide information on metabolic capabilities of the microorganisms.

It is known that methanogens are difficult for cultivation and isolation. Hence, information from 16S rRNA gene sequence analyses is particularly important, especially because it allows characterizing metabolic properties of microorganisms. Population structures were studied by using polymerase chain reaction (PCR) amplification of microbial community DNA with primers specific for archaeal 16S rRNA genes. Clone libraries of PCR amplicons from methanogenic enrichments were constructed for partial sequencing and phylogenetic analysis.

The PCR-amplified 16S rRNA sequences from methanogenic enrichments of the Dagang oil field had demonstrated 97.3 to 100% similarity to those of cultured methanogens sequences deposited in the RDP. All the sequences (101/101) were clustered in the order *Methanobacteriales* (*Methanothermobacter thermoautotrophicus* and *Methanothermobacter spp.*), which contains methanogens utilizing H_2/CO_2 and sometimes formate or alcohols as substrate for methanogenesis [Whitman et al., 1992].

The archaeal methanogenic sequences from the Dagang oil field exhibited less diversity than that reported in previous petroleum reservoir studies [Orphan et al., 2000, 2003]. This low population diversity was obtained from a relatively big methanogenic clone library consisting of 101 clones. Low diversity is not unusual for methanogen libraries and has been reported for Archaea in another anaerobic habitat [Hinrichs et al., 1999]. In deep subsurface sediments from the Nankai Through the *Methanobacteriales* were the most active methanogenic population and the dominance of H_2/CO_2-based methanogenesis was demonstrated [Newberry et al., 2004]. In saline subsurface oil field environments, hydrogenotrophic and methylotrophic, but not aceticlastic methanogens were present [Ollivier et al., 1998]. In a low-temperature petroleum hydrocarbon-contaminated aquifer, both hydrogen-consuming and aceticlastic methanogens were involved in the terminal step of hydrocarbon degradation [Kleikemper et al., 2005].

Up to now, all pure cultures of methanogens isolated by different researchers from geographically remote high-temperature petroleum reservoirs were hydrogenotrophic methanogens of the genera *Methanothermobacter*, *Methanococcus* and *Methanoculleus* [Magot et al., 2000]. Microbial assemblages from the high-temperature oil fields of Alaska (Kuparuk) and North Sea (Ninian) also produced methane from H_2+CO_2, but not from acetate [Mueller, Nielsen, 1996]. Acetate was present in North Sea oil field reservoirs at concentrations up to 20 mM [Barth, Riis, 1992], and methane production was detected in enrichment cultures with acetate as substrate at 70, 80 and 92 °C, but pure cultures growing on acetate were not isolated [Nilsen, Torsvik, 1996]. Hydrogenotrophic methanogens dominated in the Samotlor oil field also, but acetoclastic methanogens were not found [Bonch-Osmolovskaya et al., 2003].

In the Californian petroleum reservoirs, Orphan and coauthers [2000] succeeded in obtaining 16S rRNA sequences of lithotrophic methanogens of the species *M. thermautotrophicus, Methanococcus thermolithotrophicus, Methanoculleus* sp., and a 16S rRNA sequence of a new acetoclastic methanogen distantly related to the known cultured relative *Methanosarcina thermophila* (<88% similarity). Such an organism might be responsible for hyperthermophilic acetoclastic methanogenesis in oil reservoirs.

In the Dagang oil field methane was produced in the aerobic-anaerobic trophic chain based on petroleum biotransformation. The obtaining of phylotypes of the genus *Methanothermobacter* allows us to conclude that hydrogenotrophic methanogens predominated in the studied thermophilic subsurface environment. The absence of phylotypes of acetotrophic methanogens in the Dagang oil field and results of investigation of thermophilic methanogens from the oil fields obtained by other authors [Nilsen, Torsvik, 1996; Davydova-Charakhch'yan et al., 1992] allow us to suggest that acetate is degraded indirectly by syntrophic cultures, rather than directly by acetoclastic methanogens.

Acknowledgements

This work was supported by grants from Russian Foundation for Basic Research (project numbers 01-04-49250, 02-04-39002), from Schlumberger Research and Development Inc. (CRDF project No. RB0-1364-MO-02), from Russian Ministry of Science and Technology (Russian-Chinese project) and from China National Petroleum Corporation (Contract for carrying out the pilot tests of the oil recovery enhancement biotechnology [MEOR – Dagang]).

References

[1] T. Barth, M. Riis: *Org. Geochem*. 19, 455 (1992).
[2] S. S. Belyaev, I. A. Borzekov: In: *Biogeochemistry of global change*. Chapman and Hall. NY-London, 1993, 825.
[3] S. S. Belyaev, M. V. Ivanov: *Microbiology* (English translation of *Mikrobiologiya*), **44**, 166 (1975).

[4] S. S. Belyaev, R. Wolkin, W. R. Kenealy, M. J. Deniro, S. Epstein, J. G. Zeikus: *Appl. Environ. Microbiol.*, **45**, 691 (1983).

[5] S. S. Belyaev, A. Y. Obraztsova, K. S. Laurinavichus, L. V. Bezrukova: *Microbiology* (English translation of *Mikrobiologiya*), **55**, 821 (1986).

[6] H. C. Birnboim: *Meth. Enzymol.*, **100**, 243 (1983).

[7] E. A. Bonch-Osmolovskaya, M. L. Miroshnichenko, A. V. Lebedinsky, et al: *Appl. Environ. Microbiol.*, **69**(10), 6143 (2003).

[8] C. F. Brunk, E. Avaniss-Aghajani, C. A. Brunk: *Appl. Environ. Microbiol.*, **61**, 872 (1996).

[9] A. Davidova, H. J. M. Harmsen, A. J. M. Stams, S. S. Belyaev, A. J. B. Zehnder: *Antonie van Leeuwenhoek*, **71**, 313 (1997).

[10] A. Davydova-Charakhch'yan, V. G. Kuznetsova, L. L. Mityushina, S. S. Belyaev: *Microbiology* (English translation of *Mikrobiologiya*), **61**, 202 (1992).

[11] E. F. Delong: *Proc. Natl. Acad. Sci. USA*, **80**, 5685 (1992).

[12] T. H. Jukes, C. R. Cantor: In *Mammalian Protein Metabolism*, vol. 3, pp. 21. Edited by H. N. Munro. New York: Academic Press (1969).

[13] K. U. Hinrichs, J. M. Hayes, S. P. Sylva, P. G. Brewer, E. F. Delong: *Nature*. 398, 802 (1999).

[14] R. E. Hungate: In: Methods in Microbiology (Norris J.L. and Ribbons D.W., Eds.), Vol. 3b, pp. 117. Academic Press, New York (1969).

[15] M.V. Ivanov, L. S. Terebkova: *Mikrobiologiya* (In Russian), **28**, 251 (1959).

[16] J. Kleikemper, S. A. Pombo, M. H. Schroth, W. V. Sigler, M. Pesaro, J. Zeyer: *Appl. Environ. Microbiol*, **71**, 149 (2005).

[17] L. Koch: In: *Methods for General and Molecular Bacteriology* (Gerhardt, P., et al., Eds.), pp. 257. American Society for Microbiology, Washington, DC, 1994.

[18] T. V. Kolganova, B. B. Kuznetsov, T. P. Tourova: *Microbiology* (English translation of *Mikrobiologiya*), **71**, 283 (2002).

[19] S. V. Kotelnikova, A. Y. Obraztsova, K.-H. Blotevogel, I. N. Popov: *Int. J. Syst. Bacteriol.*, **43**, 591 (1993).

[20] M. Kurr, R. Huber, H. Konig, H. W. Jannasch, H. Frike, A. Trinkone, J. K. Kristjansson, K. O. Stetter: *Arch. Microbiol.*, **156**, 239 (1991).

[21] S. I. Kuznetsov: *Mikrobiologiya* (In Russian), **19**, 193 (1950).

[22] K. S. Laurinavichus, S. S. Belyaev: *Microbiology* (English translation of *Mikrobiologiya*), **47**, 1115 (1978).

[23] J. Y. Leu, C. P. Mcgovern-Traa, A. J. R. Porter, W. J. Harris, W. A. Hamilton: *Anaerobe*, **4**, 165 (1998).

[24] B. L. Maidak, G. J. Olsen, N. Larsen, R. Overbeek, M. J. Mccaughey, C. R. Woese: *Nucleic Acids Res.*, **25**, 109 (1997).

[25] M. Magot, B. Ollivier, B. K. C. Patel: *Antonie van Leeuwenhoeck*, **77**, 103 (2000).

[26] R. F. Mueller, P. H. Nielsen: *Appl. Environ. Microbiol.*, **62**, 3083 (1996).

[27] T. N. Nazina: *Mikrobiologiya* (In Russian), **53**(1), 149 (1984).

[28] T. N. Nazina, E. P. Rozanova: *Microbiology* (English translation of *Mikrobiologiya*), **49**, 104 (1980).

[29] T. N. Nazina, E. P. Rozanova, S. I. Kuznetsov: *Geomicrobiol. J.*, **4**, 103 (1985).

[30] T. N. Nazina, A. E. Ivanova, I. A. Borzekov, S. S. Belyaev, M. V. Ivanov: *Geomicrobiol. J.*, **13**, 181 (1995).

[31] T. N. Nazina, Y.-F. Xue, X.-Y. Wang, S. S. Belyaev, M. V. Ivanov: *Resource and Environ. Biotechnol.*, **3**, 109 (2000).

[32] C. J. Newberry, G. Webster, B. A. Cragg, R. J. Parkes, A. J. Weghtman, J. C. Fry: *Environ. Microbiol.*, **6**(3), 274 (2004).

[33] S. Ni, C. R. Woese, H. C. Aldrich, D. R. Boone: *Int. J. Syst. Bacteriol.,* **44**, 357 (1994).

[34] R. K. Nilsen, T. Torsvik: *Appl. Environ. Microbiol.*, **62**, 728 (1996).

[35] T. K. Ng, P. Weimer, L. J. Gawel: *Geomicrobiol. J.*, **7**, 85 (1989).

[36] Y. Obraztsova, O. V. Shipin, L. V. Bezrukova, S. S. Belyaev: *Microbiology* (English translation of *Mikrobiologiya*), **56**, 523 (1987a).

[37] Y. Obraztsova, A.Y., V. E. Tzyban, K. S. Laurinavichus, L. V. Bezrukova, S. S. Belyaev: *Microbiology* (English translation of *Mikrobiologiya*), **56**, 807 (1987b).

[38] B. Ollivier, M.-L. Fardeau, J.-L. Cayol, M. Magot, B. K. C. Patel, G. Prensier, J.-L. Garcia: *Int. J. Syst. Bacteriol.*, **48**, 821 (1998).

[39] T. C. Onstott, T. J. Phelps, F. S. Colwell et al.: *Geomicrobiol. J.*, **15**, 353 (1998).

[40] V. J. Orphan, L. T. Taylor, D. Hafenbradl, E. F. DeLONG: *Appl. Environ. Microbiol.*, **66**, 700 (2000).

[41] V. J. Orphan, S. K. Goffredi, E. F. Delong, J. R. Boles: *Geomicrobiol. J.,* **20**, 295 (2003).

[42] J. R. Postgate: *The sulfate-reducing bacteria.* 2nd ed. Cambridge Univ. Press. Cambridge, (1984).

[43] A. Reznikov, E. P. Mulikovskaya, I. Yu. Sokolov: *Metody analiza prirodnykh vod* (*Analytical Methods for Natural Waters*), Nedra, Moscow, 1970.

[44] E. P. Rozanova, A. S. Savvichev, S. G. Karavaiko, Yu. M. Miller: *Microbiology* (English translation of *Mikrobiologiya*), **64**, 85 (1995).

[45] H. G. Truper and H. G. Schlegel: *Antonie van Leeuwenhoek*, **30**, 225 (1964).

[46] G. Voordouw, S. M. Armstrong, M. F. Reimer, B. Fouts, A. J. Telang, Y. Shen, D. Gevertz: *Appl. Environ. Microbiol.*, **62**, 1623 (1996).

[47] W. B. Whitman, T. L. Bowen, D. R. Boone: In: *The Prokaryotes* (Starr, M.P., et al., Eds.), Springer Verlag, Berlin, pp. 3352, (1992).

[48] E. A. Wolin, M. J. Wolin, R. S. Wolfe: *J. Biol. Chem.*, **238**, 2882 (1963).

[49] J. G. Zeikus, P. J. Weimer, D. R. Nelson, L. Daniels: *Arch. Microbiol.*, **104**, 129 (1975).

INDEX

A

B

C

P

R

S